More Praise for *Righteous Porkchop*

"The story that Nicolette Hahn Niman tells in this book is full of heroes and villains (of the two-footed kind). Food lovers can only hope that America takes her message to heart and votes at the checkout counter."

—Peter Kaminsky, author of *Pig Perfect:*
Encounters with Remarkable Swine

"Finally, a book that can help everyday Americans understand what's at stake as a result of our factory animal systems. A great, common sense and steady read."

—Michel Nischan, chef, Dressing Room;
president, Wholesome Wave Foundation

"To tell her story, Nicolette Hahn Niman has created a unique genre—half romance, half expose, narrated through her personal experience of the horrors of factory farms, mitigated by the human face of farmers and ranchers she's met in her travels across the land. With a lawyer's mind she dissects the extremes of industrial farming; with a woman's heart she learns to mitigate her own extremes to create greener pastures for the animals, her husband, and herself on the ranch she loves."

—Betty Fussell, author of *Raising Steaks:*
The Life and Times of American Beef and
The Story of Corn

RIGHTEOUS
PORKCHOP

Finding a Life and Good Food

Beyond Factory Farms

NICOLETTE HAHN NIMAN

HARPER

NEW YORK • LONDON • TORONTO • SYDNEY

HARPER

FIRST HARPER PAPERBACK PUBLISHED 2010.

Designed by Mia Risberg

The Library of Congress has catalogued the hardcover edition as follows:
Niman, Nicolette Hahn.
    Righteous porkchop : finding a life and good food beyond
    factory farms /
    by Nicolette Hahn Niman.
            p. cm.
    Includes bibliographical references.
    ISBN 978-0-06-146649-6
1. Animal waste—Environmental aspects.   2. Swine—Manure—Environmental aspects.   3. Meat industry and trade—Environmental aspects.   4. Animal welfare.   I. Title.
II. Title: Finding a life and good food beyond factory farms.
TD930.2.N56 2009
636.08' 32-dc22   2008027852

ISBN 978-0-06-199845-4 (pbk.)

10 11 12 13 14   OV/RRD   10 9 8 7 6 5 4 3 2 1

For my parents, Mutti and Vati.
By word and example, you taught me
how to express myself and to
respect every living thing.

CONTENTS

FOREWORD

I first met Nicolette Hahn Niman in 2000 when she was working as a lawyer for National Wildlife Federation. Inspired by her commitment to public service and her political and legal talent, I offered her the job as staff attorney to the newly established Waterkeeper Alliance. Soon after she came to work in our cramped basement offices in White Plains, New York, I asked her to head up a national campaign to combat hog factories. She seemed particularly suited for this job since she had formerly served as a county prosecutor in North Carolina, the epicenter of industrial pork factories. She threw herself into the fight with energy and passion; she seemed to have found her calling. Using her well-honed skills as a litigator and an inborn talent for grassroots organizing, Nicolette quickly made herself one of the nation's leading warriors and strategists in the apocalyptical battle against factory meat production.

By then the factory meat industry was already vying for the blue ribbon as America's worst polluter. As is often true with large scale pollution, the destruction of public resources accompanied the subversion of democracy. Factory meat barons were dominating the market place, not through efficiencies, as they pretended, but by ruthless market control and by externalizing their pollution costs. The big integrators could not produce porkchops or bacon cheaper than traditional family farmers unless they got around environmental laws. Since the industry produced far more waste than it could profitably dispose of, its entire business plan was based on its capacity to illegally dump industrial scale quantities of raw animal waste and toxic chemicals into the nation's waters and air and get away with it! To take advantage of public good will toward small farmers, big agribusiness employed slick PR firms and lobbyists to defend animal factories by costuming them as Old MacDonald's farm. Using strategically orchestrated public deception, intimidation, campaign contributions, and political clout, meat moguls captured state and federal environmental agencies which obligingly (and illegally) waived environmental permit requirements, routinely ignored flagrant violations, and transformed themselves into taxpayer funded hand puppets aggressively defending industry mischief. With the exception of coal mining's mountaintop removal, no other industry in America's modern history had gotten away with systematic pollution of this magnitude.

North Carolina was especially hard hit by big animal operations. Just a few years earlier, the state had been famous for some of the purest rivers and coastal waters in the United States. But all that had changed. We saw rivers and streams choked with algae fed by nutrient discharges from hog and poultry factories. Manure contamination caused closures of waterways and beaches, and devastated fisheries. In 1991, more than a billion fish died in the Neuse River, their bodies covered with open bleeding lesions. Bulldozers were used to clear the resulting mountains of dead fish from public beaches. In 1995, a spill from one lagoon killed millions more fish in North Carolina's New River. Since then industrial pork producers have contributed to the deaths of millions of fish through contamination of North Carolina's public waters. Arguing that animal factories are one of the most pressing environmental and social issues of our era, our Neuse River keeper, Rick Dove, urged me, with irresistible persistence, to challenge the industry on a national level. Nicolette took on the job.

We regarded Smithfield Foods as the chief villain in North Carolina. Smithfield was once strictly a meatpacker with no involvement in owning pig farms. But then its CEO, Joe Luter, began buying up farms to control all aspects of pork production—"from squeal to meal," the company boasted. Luter is known for a ruthless approach to business that maximizes Smithfield's short-term profits, seeming to have little regard for the long term costs to communities. Smithfield's takeover of North Carolina's pig farming entrenched a harsh system of meat production that eliminates both animal husbandry and family farmers and subjects livestock and workers to unspeakable conditions.

Smithfield's system ignores both civilized norms and legal niceties. Indeed, based on its own records, the company can be fairly characterized as a chronic lawbreaker. Its facilities across the country are regularly cited for violating various environmental and labor laws. In 1998 alone, Smithfield was fined $12.6 million for 6,900 violations of the federal Clean Water Act, the largest penalty in the statute's history.

The waste from hog factories is prodigious. A hog facility with 100,000 animals can produce the same amount of fecal waste as a city of one million people. (One of Smithfield's Utah factories, Circle Four Farms, houses 850,000 pigs and produces more fecal waste than the people of New York City.) Hog factory waste is stored in cesspools romantically called lagoons. While cities must treat sewage before discharging it, confinement facilities spray their lagoon waste untreated onto fields, which quickly become saturated. The lagoons themselves often leak or overflow. The toxic soup then seeps into groundwater or is carried by rain into nearby streams or lakes. Waste from these factories can contain a witch's brew of nearly 400 dangerous substances—including heavy metals, antibiotics, biocides, chemical disinfectants, pesticides and disease-causing viruses and microbes. Antibiotic residues in this lethal soup foster the growth of deadly 'super bugs'—disease organisms that are immune to human antibiotics.

Millions of tons of fecal stew produced by hog factories have poisoned the drinking water of scores of rural citizens with deadly nitrates that can cause miscarriages in pregnant women and blue baby syndrome in infants.

Animal factories also produce dangerous and noxious odors. The putrid stenches from hog operations can defy description. I've met farmers living near hog facilities who have choked, vomited and fainted from the fetid gases as they worked their fields. The smell cannot be removed from skin or clothing—even with the strongest soaps. Food eaten even a mile downwind of a pig factory can take on the odor and flavor of pig feces. In summer, downwind neighbors are often unable to sit on their porches, open their windows, hang their laundry outside, or enjoy their meals. The odors can be so intense that they have been known to nauseate people flying in airplanes three thousand feet above the facilities.

The growth of meat factories is the principal factor in the decline of the American family farm. Each pig factory puts family farmers out of business, replacing high-quality agricultural jobs with hourly–wage workers in degrading positions that are among the lowest-paid and most dangerous in the United States. Because the animals are fed and watered by computer and are given almost no husbandry, as few as two workers may tend an operation with ten thousand pigs. Conditions are so miserable that employees seldom endure these jobs for more than a few months. Major slaughterhouses, including those owned by Smithfield, typically have a 100 percent annual employee turnover rate.

The meat industry uses its money, power and political clout to silence the critics and disable democracy in order to get away with its bad conduct. In 1996 the *Raleigh News & Observer* won the Pulitzer Prize for its "Boss Hog" series which showed how the hog industry had corrupted North Carolina politicians to grease the skids for its illegal activities. Using its proxies, the American

Farm Bureau and the Pork Producers Council, the meat moguls have influenced state legislators across the country into ignoring their oath to safeguard the constitutional guarantees of speech and debate and the democratic tradition of open government. State legislatures under the sway of agribusiness have passed laws making it illegal to photograph factory farms. Eleven states have enacted laws making it illegal to publicly criticize factory farm food. Oprah Winfrey endured a two-month jury trial in Amarillo, Texas, after hosting a show in which a guest questioned the safety of industrially produced meat.

The industry routinely flexes its political muscle to muzzle government scientists and prohibit them from publishing studies that show the significant public health threats posed by industrial agribusiness. Nicolette and I have personally experienced the industry's efforts to stifle debate as we've travelled the country to speak against factory farms and prosecute the industry for its lawbreaking activities in various states.

For years the industry assigned a man to follow me around the country, sit in the front row at my speeches and press events and pointedly tape record all my remarks. He would follow me in hallways or dark parking lots after I spoke, often making nasty remarks. He shadowed me from California to New York and dozens of states in between. He maintained a Web site tracking my whereabouts and exposing my "lies." I was only relieved of his constant company after he was convicted of "theft by deception" in Nebraska. His court ordered probation prevented him from leaving the state.

Smithfield recently filed criminal charges against me in Poland for speaking out against its destructive style of meat production there. The company never dared sue me in the United States where the First Amendment guarantees of freedom of the press have set a high bar for libel actions. In Poland, however, where press freedom is a less sanctified tradition, Smithfield is currently suing me for remarks I made during a debate with Smithfield's Polish director before the Polish Parliament. While U.S. law makes truth an absolute defense to libel, the Polish law criminalizes insulting remarks against corporations even when they are accurate. So I now find myself facing a possible two-year prison sentence in that country. Smithfield thrives best in venues where democracy is hobbled or incomplete.

In *Righteous Porkchop*, Nicolette eloquently and vividly describes the evils of industrial animal agriculture not just for the environment, but for people and animals. She also lays out why industrial production is a threat to our most treasured traditions and values, even our democracy. Most significantly, she demonstrates convincingly that factory meat production, with all its attendant pollution, cruelty, and economic and social disruptions, is unnecessary; we can choose to raise farm animals in ways that are environmentally sound, healthful, humane, and consistent with America's historic mission as an exemplary nation.

—*Robert F. Kennedy, Jr.*

CHAPTER ONE

My Crash Course in Modern Meat

A NEW ASSIGNMENT

There I was, driving through sheets of relentless rain, straining to get a good view of the road in front of me. The year was 2000 and I was heading east on I-80 toward my new job just north of New York City, anxiously anticipating what lay ahead. As I made the trip from Michigan, I thought about all I'd given up. I'd just quit my job, sold my house, given away most of my possessions, and crammed the rest—my least expendable belongings—into my aging Volkswagen Golf. I had the overwhelming sense that this new beginning would be a turning point in my life. A few days later I would start as the senior attorney for the environmental group, Waterkeeper, headed by Robert F. Kennedy, Jr.

After some long hours on the road, I finally reached New York and collapsed on a friend's sofa. Two days later, my first day on the new job, I settled into a small, austere office at the Pace University

Law School, where Waterkeeper was housed, and awaited direction. These initial weeks on the job turned out to be atypical in both the tasks I was given and the hours I worked. Bobby (as everyone calls him) threw a hodgepodge of requests at me, relating to everything from employment issues, to air pollution, to the organization's tax status. These were tests, I suspected, yet none of the work seemed particularly pressing. Diligently but dispassionately, I plodded along as a legal factotum in a regular rhythm of nine to five.

That all changed one Saturday afternoon when I got a call from Bobby (for whom a mere wisp of a line separates on- and off-duty). "Nicolette, I want you to take charge of our hog campaign," he barked in a way that sounded half command, half request. "You'll have a lot of autonomy and responsibility," he continued, "but it's also going to be a lot of hard work." At that moment, there was really no "hog campaign" to speak of. It was little more than Bobby's notion that he wanted to sue hog farming operations for contaminating rivers with their manure and that he wanted it to be part of a larger national crusade against industrialized animal operations that caused pollution.

The responsibility and autonomy were certainly appealing, but I knew almost nothing about hog farming and it struck me as, well, an immersion in poop. It was not exactly the glamorous job I'd envisioned when abandoning everything for New York. "Uh—I'm not sure I want to work full-time on manure," I ventured.

There was another reason for my reticence. I'd sought this job with the idea of dedicating myself to environmental causes dear to my heart, yet livestock farming didn't hold much interest for me. Just after my freshman year of college, a tangle of vaguely informed concerns about the environment, health, and animals had inspired me to quit meat. However, since I wasn't much of a proselytizing vegetarian, I'd largely ignored the dark details lately emerging about the meat industry. Frankly, I found those stories so depressing I intentionally avoided them. (Anyway, why did I need to read that stuff—wasn't I doing my part by abstaining from meat?)

Hearing my hesitation, Bobby responded that before giving my answer, I should see for myself what this was really about. "Just go to Missouri and meet the people who've been asking for our help. Then you can decide."

A STEAMROLLED COMMUNITY

As Bobby is not a person easily gainsaid, a few days later I found myself stepping off a plane in Kansas City. My ultimate destination was a Missouri town three hours to the northeast that had become densely populated by the hog operations of the large agribusiness corporation, Premium Standard Farms (PSF). Two farmers, a lawyer, and an environmental advocate would be my guides. They met me at the airport terminal exit in a white rental van.

From our phone conversations, I already knew that Scott Dye, a Sierra Club employee in the group, was a straight-talking, salty-tongued fountain of knowledge. With his grizzled beard, booming voice, red plaid shirt, and baseball cap, he struck me as more lumberjack than tree hugger. Scott gave my hand a firm shake then introduced me to the others.

From the introductions I learned that all but one in the group came from farming families. That fact stood out to me because I had already encountered claims from agribusiness that complaints about industrial animal operations are made only by misguided "displaced city-dwellers" who simply don't understand agriculture. It seemed to be a classic agribusiness response to any criticism.

As we drove north, I heard facts and stories about the people who'd been raising crops and animals in the area for generations, long before big agribusiness moved in. Seated next to me was one of them: Terry Spence, a second-generation farmer with a Red Angus cattle herd. He seemed a modest, soft spoken man. But I soon detected a will of iron underneath as he described the company's inauspicious arrival in their town a decade earlier.

At the time, Terry was serving on the local township board. When he and his fellow board members heard that PSF planned to move to their community after being blacklisted in neighboring Iowa, they leapt into action. The board drafted land use ordinances that would prohibit animal operations from causing pollution and odor, laws by which traditional farmers could easily abide.

The company responded by making a beeline to the state capitol to flex its political muscle. The day after the township adopted its anti-pollution ordinance, the state approved PSF's permits, effectively overriding the local laws. The legislature even passed a law that explicitly exempted Terry's county from the protection of a decades-old state statute that makes farming by out-of-state corporations illegal. Everything coming out of the capitol appeared hand-tailored for the company to seamlessly set up its facilities in Terry's township.

PSF announced plans to erect ninety-six buildings that would hold more than one-hundred thousand hogs within a one-mile radius of more than twenty family homes. The rise of visible, vocal community opposition did little to slow the commencement of construction.

For good measure, PSF sued the township and its board, apparently hoping to intimidate them into repealing their ordinances. But the tiny community stood fast. "The company underestimated how much pride we had in our way of life, and how determined we were to protect it," Terry wistfully recounted. When PSF won the case, the township appealed the decision all they way to the Missouri Supreme Court. Still, it got no relief. The state's highest court ruled that the board was powerless to regulate local agricultural buildings and manure storage systems. The battle to protect Terry's hometown, begun almost ten years before I met him and still raging, clearly required remarkable resilience and tenacity.

Terry's tale was jarring to me both as a lawyer and a former city council member. Local governments in the United States have always held firm control over local land use, under a principle called "home rule." The Missouri courts and legislature were essentially abandoning a centuries-old system of self-governance.

Sitting in the van chatting with Terry, Scott, and the others, I was drawn into their saga and feeling entirely comfortable. Then we stopped for lunch. I knew that with their rural farming backgrounds, several of my companions would have been exposed to years of meat industry propaganda that equated vegetarians with bomb-throwing radicals plotting to overthrow the American way of life. I discretely ordered an egg salad sandwich. When it seemed to pass unnoticed, I breathed a small sigh of relief.

Across the lunch table from me was the lawyer in the group, Charlie Speer, an amiable, scholarly looking man with silver rimmed glasses and receding gray hair. Charlie explained that he'd been representing Terry and several neighbors against PSF in a lawsuit over water pollution and odor. They'd had a tough row to hoe because as part of its unstinting generosity toward PSF, the state legislature had adopted a law making it almost impossible to sue hog facilities for nuisance.

I was stunned. In law school I'd been taught that hundreds of years of American and English case law hold that no person can use his property in a way that causes a nuisance for his neighbors. This age-old precedent was apparently being entirely ignored. "They call the nuisance exception the 'Right to Farm' law, but it's got nothing to do with real farmers," said Charlie, shaking his head. "Let's just call that what it is—a bunch of *horseshit*," interjected Scott in hearty agreement. "I mean, my clients are all farmers," Charlie continued, "and they're the ones hurt by the law. It's really to protect a few big companies, like PSF. The legislature is just giving them a free pass." The genuine anger coloring Charlie's cheeks and voice, I suspected, was what kept him motivated in spite of what seemed very unpromising odds.

FACTORIES, NOT FARMS

When we reached Putnam County, Scott drove straight to an enclave of large windowless metal warehouses. "There are seventy-two of 'em at this point," Scott narrated unhappily. I could picture furniture or salt mounds stored in the barren structures, but not living creatures. More than a thousand pigs were crowded into each building, where they'd spend every day in pens with concrete floors, my guides explained.

The floors were slatted so the pig manure and urine would collect in containments below the buildings. By standing and lying in their own excrement, pigs push it through the slats with their hooves and bodies. As with toilets, water is added to the waste for ease of transport. This liquefied manure then flows out to large, open-air storage ponds. When the ponds reach capacity, the waste is pumped into giant wagons that spread it onto surrounding farm land. This elaborate mechanized waste handling system, my guides pointed out, makes it possible to keep huge numbers of animals in buildings, with very few people looking after them.

Perusing the landscape, I spotted numerous giant ponds beyond the buildings, roughly the size and shape of football fields. "They call those 'lagoons,'" Scott said with a skeptical snort. It did seem an odd word for a murky brown manure hole bordered with weeds and scum. Until that moment, I'd always pictured a "lagoon" as a crystal clear azure tropical pool surrounded by sandy white beaches.

My brow furrowed as I studied the metal sheds and absorbed the description of the grim drama unfolding inside, shielded from our view. Aside from a couple of petting zoos, the only pigs I'd ever seen in the flesh were wild boars rooting and roaming the forests of Germany. It was hard to imagine their descendants surviving in lifeless buildings like these. "You mean the animals *never* go outside?" I questioned tentatively. "Never," Scott called over his shoulder from the driver's seat. "They're born in confinement and they're there until they take 'em to slaughter." "So, I guess it's sort of like living in a barn," I reasoned. "No, it's not like a barn," Scott patiently explained to the naïve city dweller. "You see, Premium Standard Farms never even gives them any straw to lie on. That would be like throwing a handful of straw in your toilet—it would gum up their fancy sewer system." Just talking about it seemed to exasperate Scott. "Listen—these aren't farms. They're *factories*."

The insides of confinement buildings were organized as two rows of crowded pens, explained Scott. A pig's opportunity for movement was minimal; exercise was nonexistent. Eating was the sole activity. Night and day, the animals languished with nothing to do but stand and lie in their dirty pens. The sows, females used for breeding, had it the worst. They were continually contained in individual metal cages so narrow they could not even turn around. This was all starting to sound like a pig prison, and it was getting me depressed.

As our van rolled slowly past row after row of gray pig barracks, my eyes scanned the landscapes, which felt eerily devoid of life. I was being told that tens of thousands of pigs were just yards away, yet not a single animal or person was anywhere in sight. Everything about the place felt horribly wrong.

It began dawning on me then that perhaps divorcing animals from nature is exactly the point of the industrial approach. By taking animals off the land and placing them in totally artificial settings, the operations sever themselves from sunlight, seasons, weather, and even the need for skilled people with an understanding of animals. When I was growing up, farmers were the folks my father had always chatted with about local rainfall, heat, and wind. But none of that mattered here. Everything was being done inside buildings and by machines, much of it entirely automated. Operators could spend more time staring at their gauges and spreadsheets than with the animals in their charge. I suspected they did. This did not even remotely resemble my idea of "farming."

After cruising the area for about twenty minutes, Scott pulled the van off to the shoulder. He'd been running the air conditioner on the re-circulate mode and now felt that I should experience what it was really like outside. "Let's get out and take a whiff," he suggested sardonically while shutting off the motor. Our group piled out at the roadside into a sticky mid-afternoon sun. Heat and odor instantly pressed in on me from all sides. A mildly nauseating smell of rotten eggs enveloped us. "Hmm. Hydrogen sulfide," I speculated. "Yup. Hydrogen sulfide and ammonia, and plenty of it," Scott confirmed with a nod.

Industrial animal operations, especially hog facilities, are notorious for foul odors. I'd been reading and hearing that the stench could be unbearably oppressive. But it wasn't as strong as I'd anticipated, especially considering we were quite close to a few of the buildings. This was unpleasant, but not intolerable. Perhaps reading my thoughts, Charlie remarked, "The smell gets a lot worse than this. A *lot* worse, depending on the temperature, humidity . . . wind. Actually, the most frustrating part for neighbors is that they just never know when it's going to get really bad."

Standing by the roadside I could now *hear* the hog operations, too. The air was vibrating with a whirring sound, reminiscent of an airplane revving its engines for takeoff. Pointing to a huge fan at the end of one of the buildings, Scott informed me I was hearing the ventilation systems. "The fans have to run 24/7 because of the fumes inside." If the ventilators stop working, even for a few minutes, he explained, tragedy strikes. "Every year we hear about incidents in the area where power goes out for a while and all the pigs suffocate. It doesn't take but a few hours." It occurred to me that such toxic air couldn't be good for the people working there, either, even when the fans were working properly. I'd never imagined that a "farm" could be such a hazardous place for man and beast.

We all climbed back into the van and Scott pointed it toward the farm where he grew up, less than two miles from the spot where we'd been standing. On the way, Scott explained that since his father's death a tenant farmer cultivated his family's crop land. His mother was still living in the family home. We pulled up the long gravel driveway of a white 1930s clapboard farmhouse with an empty clothesline out back. A row of PSF's metal buildings glimmered ominously in the distance.

A white-haired woman warmly greeted us at the door, inviting us all in and asking if we'd care for some cold lemonade. I gladly accepted the offer as we took seats in the living room. Mrs. Dye served us each our drinks then sat down. "Mom, tell Nicolette what's it's been like livin' here since PSF came to town," Scott gently prodded. His mother nodded and slowly began talking. "Well, it's just made it hard to live here," she began.

She told me how she now struggled to enjoy living in her home, where she'd raised her children and passed most of her life. Mrs. Dye suffered under the constant uncertainty of when PSF's stench would next invade her yard and house. Some days it even affected her breathing. "I don't care to hang my laundry outside anymore," she told us with a sigh, "because when I bring it in a lot of times it smells like hog manure." Because nighttime was when the fumes were strongest, even a good sleep was elusive.

As we said goodbye to Mrs. Dye, I watched Scott's hulking body wrap his frail mother in a bear hug. Now I knew why Scott was so dedicated to fighting hog factories—he's a good son.

As we drove away, the other farmer in our group piped up. For him and his family, too, he said, the nasty smell was the worst part. Rolf Christen, an organic crop, cattle, and free-range chicken farmer originally from Switzerland, had moved to Missouri with his wife almost two decades earlier to start a diversified family farm. About ten years after they bought their land, the PSF operations set up shop. Like Terry, he'd spent the years since tangling with the company. What was most upsetting, he said, was how it disrupted the lives of his family members. Sometimes he and his wife and children would wake up in the middle of the night as a putrid smell rolled into their house like a dense fog. "We rush around closing all the windows but it doesn't really do much to keep out the odor," he recounted with dismay.

Our final stop was along a stream that snaked through the PSF operations. Terry was part of the local citizens group, Missouri Stream Team 714, that monitored water quality. "For almost ten years, our team took water samples here every month," Terry explained, "and the nitrate levels just kept going up and up." I already knew that the nitrates contained in manure are the main water pollution problem from industrialized livestock operations. Every gardener is familiar with nitrogen's value in making plants grow. But the same qualities that make it good fertilizer render it hazardous to aquatic plants and animals. When too much nitrogen gets into the water, algae bloom out of control, depleting all the oxygen and causing just about everything else to die (a condition called "hypoxia").

Terry and fellow team members had dutifully made the government and PSF aware of their sampling results. They then spent years trying to get them to respond to the rising nitrate levels. The state did essentially nothing. The company, however, did take some action, although not to stop the pollution. "The most ridiculous thing is that the only thing PSF did in response was to put that up," he bitterly complained as he pointed to a chain-link fence just below where we were standing. "Now I can't take my samples." The new fence cut off the path, making it impossible to step down to the stream.

This was all so disconcerting. How could our government refuse to act when a corporation was wreaking such havoc on a community of hardworking, salt of the earth citizens, and perhaps even breaking the law? It was becoming clear to me that fighting industrial animal facilities involved more than combating poop. Soon I would learn that Missouri's refusal to enforce the environmental laws was repeated everywhere across the nation. Collectively, these failures had made pollution from industrial animal operations one of the United States' most serious water pollution problems.

I already knew that manure wasn't necessarily a negative. I remembered the small farms I'd known as a child in Michigan, where a pile of manure was treated as a treasure trove. Driving back to the airport, I remarked to Scott: "So, manure isn't always bad, right?" "Hell, no!" Scott cried. "Farmers around here always fertilized their fields with manure. Environmentally, it's actually the best fertilizer. And we didn't have any serious problems with pollution. It's just now, there's too much concentration. Too many animals in one place. Way too many."

I left Missouri with a lot to consider. I'd been deeply moved by what I'd seen, but didn't want to make any promises. "Let's talk very soon," I suggested to Scott as he dropped me off at the airport. On the plane back to New York, the day replayed itself in my head. We were being asked, as an environmental organization, to target the air and water pollution generated by industrial hog facilities. Their polluting potential was becoming patently obvious. But what was already haunting me more was how these operations directly affected people and animals.

I'd now had my first up close glimpse of our modern, industrial food system, and I didn't like what I was seeing. I was witnessing how corporations have been quietly but radically altering how our food is produced. It's a shift away from farming as a true profession, passed down from generation to generation, and carried out by skilled people closely connected with their animals, lands, and communities. In its place agribusiness has been substituting factory-style industrial production, mechanized, automated, and carried out by hourly laborers who often have no training in animal husbandry and no connections to the surrounding community.

ORIENTATION, PART TWO

The next day, when I returned to the office, Bobby—an intensely action-oriented man—grilled me impatiently on when I'd be getting our lawsuits going. But I knew I wasn't ready to start suing anyone just yet—first, I wanted to do more investigating. Industrial animal facilities were clearly reprehensible, but could we prove that they were actually breaking the law? That day I started trying to find out. I began calling and emailing questions to experts around the country—everyone from aquatic ecology professors to animal welfare advocates. And I hastily arranged a trip to North Carolina, birthplace of the corporate hog confinement system and lately the nation's second largest pork producer.

Perhaps more than any other force, industrial hog and poultry production have molded North Carolina's recent history. For most of the twentieth century, the state's many farmers raised a variety of crops and a steady number of hogs—fewer than two million—on small family farms peppering the state. Almost every farmhouse had a few pigs out back. At the time, North Carolina was famous for its crystal clear flowing streams and abundant fisheries and craberies.

However, the state's economy and physical environment began undergoing major alterations in the late-1980s. Catalyzed largely by a single entrepreneur, pig production in North Carolina exploded: quadrupling within fourteen years from 2.5 million hogs in 1989 to 10 million in 2003. During the same time period, the total number of North Carolina farms with pigs shrank from 12,500 to 2,800. Traditional farms, where pigs were raised in small, freely ranging outdoor herds, were being replaced by corporate controlled facilities with vast liquefied manure lagoons and thousands of animals living in continual confinement. Industrial chicken and turkey production were dramatically expanding as well.

Because North Carolina's natural environment was reeling from the effects of this invasion of industrial animal production, Bobby chose it as the launching site for our legal campaign. By fortunate coincidence, I was already licensed to practice law in North Carolina. Fresh out of law school, I had lived two years in the state and worked as an assistant district attorney in Durham. On this trip down there, I would review documents, interview potential witnesses, and take a firsthand look at some of the state's hog facilities. My other reason for heading to the South was to meet our point man on hog pollution.

Rick Dove, the person who'd first focused Bobby Kennedy's attention on industrial animal operations, came to pick me up at the Raleigh airport. Waterkeeper is an international alliance of water protection activists called "riverkeepers." Rick was the first riverkeeper in North Carolina. When I met him, he was a sixty-one-year-old retired Marines Corps colonel, looking every bit the part with an erect posture, cropped haircut, starched shirt, and drab olive pants tucked in polished black combat boots. Bobby had told me that

Rick had been a lawyer and judge in the Marines, and that he was a devout Catholic and a lifelong Republican who, for most of his life, would never have called himself an environmentalist. At first, much of this seemed a bit incongruous with the profile of an environmental activist. But I would soon come to understand how Rick's experiences had radicalized him. I instantly liked this interesting man, who was both serious-minded and friendly.

From our emails and phone conversations, I knew quite a bit about Rick's history. It was plain that after living two decades on its banks, Rick had a palpable intimacy with North Carolina's Neuse River. Upon retiring from the Marines, he and his son briefly tried to make a living fishing. But they were too late. The decline of North Carolina's fisheries was already in a free fall. They frequently came across fish with sores covering their bodies; shellfish and crabs were often officially declared too contaminated to eat. And when the river's pollution started to make him physically ill, Rick knew it was time to give up the fishing venture.

That experience, Rick had told me, was probably what prompted him to answer a want-ad for a "Neuse Riverkeeper" without knowing what a riverkeeper was. He just liked the sound of "keeping" the Neuse. Rick Dove soon became North Carolina's most visible environmental crusader as he redefined what it meant to be a riverkeeper.

With military precision, he organized battalions of volunteers into platoons who patrolled the river to seek out and document river contamination. They put on waders and climbed into motor boats and canoes. They snapped photos, made video recordings, scribbled in notebooks, and carefully prepared and recorded hundreds of water samples. The information Rick and his forces gathered became evidence in dozens of legal cases and negotiations to halt pollution.

Their work quickly revealed that confinement hog and poultry operations were the Neuse River's most potent enemy. When he repeatedly found himself stymied by "No Trespassing" signs near animal facilities, Rick doggedly searched out private pilots to fly him over the properties. Over time, he built a squadron of eighteen volunteer pilots.

After Rick picked me up at the Raleigh airport, we headed straight to a stash of files at the central Department of Natural Resources office (known locally as "the DNR"). By law, the state agency must monitor and record pollution at animal operations and elsewhere. The DNR files contain facility inspection reports, citizen complaints, and notices of violations cited by government agencies. The Freedom of Information Act requires that the files be made accessible to the public, including us.

Rick pulled out a yellow legal pad with a list of problem facilities he'd compiled from his own monitoring and his volunteers' observations. He already had a pretty good idea of the major offenders, which was where we'd focus our time. Each operation was under the control of Smithfield Foods, Inc. by contract or ownership. The largest pork company in the country, Smithfield controlled more than three-quarters of the state's hog raising operations and almost all of its hog slaughtering.

After a few hours of scouring the DNR papers, some of the operations floated to the top of our list. North Carolina's industrial hog facilities function very much like Missouri's, except that when liquefied manure is pumped from lagoons it's usually sprayed directly onto land with giant water cannons. The DNR files described dozens of incidents where liquefied hog waste had gotten into waterways because facilities had sprayed hog waste in the rain, manure cannons had been directed at ditches or creeks, or lagoons had burst or spilled over after being filled too high. We photocopied the records. Then we packed up, climbed into Rick's red Dodge pickup and started the two-hour drive southeast to his home in New Bern.

Over the next few days, Rick and I traveled to three regional DNR offices and pored over their files. We were covering a lot of eastern Carolina territory, and everywhere Rick pointed out dozens of hog and poultry operations along our route. No community seemed free of them. The dull gray metal sheds looked depressingly identical to each other and to the ones I'd been shown in Missouri. Some were bordered by ten-foot cyclone fences. Big signs with red lettering warned: "ABSOLUTELY NO TRESS-PASSING!" Each operation had at least one large manure lagoon nearby. Several looked abandoned. It was creepy how here, too, we never saw any people or animals.

Rick loves to laugh and, as I would learn over time, he often plays small practical jokes on his friends and colleagues. As we drove between our destinations he would occasionally tease me by rolling down the truck windows as we neared confinement facilities. There was always a rotting egg odor, which seemed strongest when manure was being aerial sprayed. But one of those times, when the wind was blowing directly at us, it wasn't all that funny. We were hit with a blast of putrid stench. A wave of nausea washed over me, and a salty liquid rushed into my mouth. I lurched toward the window and hurried to open it fearing I was about to be awfully ill. Luckily, the feeling passed without me emptying the contents of my stomach. At that moment I understood that industrial hog odors can be much more than an annoyance.

"God. The smell is sickening. It must be unbearable inside the buildings," I thought aloud after regaining my composure. "I hear more complaints about odor than anything else," Rick agreed. Bobby had talked about a riverkeeper in Ohio who'd once spent time as a laborer in a hog confinement operation. While employed there, his wife refused to let him bring his work clothes into their home. The garments, she said, stank up the house and she could not get rid of the odor. She claimed she could even smell hog manure on his breath.

To prepare for this visit to North Carolina, I had read a pile of environmental and health studies. A lot of the research related to the vapors inside hog confinement operations. It substantiated what people had been telling me—namely, that they're full of dangerous toxins. Hydrogen sulfide tops the list. A Purdue University farm safety guide emphasized that manure lagoons are inherently dangerous, soberly advising: "When animal waste is being stored in large quantities, a number of hazards are present for both man and animal." After first cautioning about the risk of *drowning* in a lagoon, the manual warns against being overcome by its gases, particularly hydrogen sulfide. The guide states: "Hydrogen sulfide (H2S) is a very poisonous gas . . . A concentration of only 50 parts per million . . . can cause dizziness, irritation of the respiratory tract, nausea, and headache. With concentrations exceeding 1,000 parts per million, death from respiratory paralysis can occur with little or no warning." Adding, to illustrate: "Recently, a 16-year-old Wisconsin farm worker collapsed and died while cleaning a confinement calf barn located above a 100,000 gallon liquid manure [pit]. Hydrogen sulfide was reported as the cause of death."

Eventually, I would uncover more than one hundred studies (mostly done in Europe, where researchers depend less on funding from agribusiness) linking the gases of industrial animal operations to various ailments, including lung diseases, nausea, nosebleeds, depression, and brain damage. Not surprisingly, the fumes are strongest inside the buildings, so workers are the most frequent human casualties of confinement fumes. The animals, who literally bathe in the vapors day and night, have it even worse. Chronic respiratory ailments are industrial hog operations' single biggest animal health problem and, as Scott had pointed out in Missouri, asphyxiations in hog confinements are commonplace.

Rick spent most of his time over the years tracking down water pollution, but began turning his attention to what was going into the air. "The hog factory guys love dismissing the fumes as just 'smells.' But that's a bunch of bullcrap. *It's air pollution*!" he railed one day as we drove to a DNR office. In fact, scientists have shown that as much as 80 percent of the nitrogen in manure lagoons ends up in the air, mostly in the form of ammonia.

In addition to lagoons, research has shown that lots of fumes are released by other parts of confined animal operations. Ammonia is continuously emitted from urine and feces in and under the hog buildings themselves, as well as anywhere liquefied manure is spread on the land. A North Carolina State University scientist and his colleagues documented that the ammonia from all aspects of hog and poultry operations was one of the state's biggest air pollution problems. The researchers reported that North Carolina's hog operations alone were putting between 55,000 and 72,000 *tons* of ammonia nitrogen into the air every year. Their research also showed the startling fact that 84 percent of the state's recent increase in airborne ammonia resulted from a single cause: hog industry expansion.

Of course, what goes up must come down. Thus, contaminants that industrial animal operations put in the air also end up polluting land and water. Research shows that as much as 60 percent of the nitrogen entering North Carolina coastal waters, including the state's fragile estuaries, comes from the atmosphere. The air to water pollution pathway has been connected to algae blooms lethal to fish and other aquatic creatures. It has also been linked to contamination of drinking water and poisoning of soils that kills trees and other plants. Some compounds released from liquefied manure storage are greenhouse gases, contributing to global warming.

In later research, I discovered that air pollution from industrial animal operations is a decidedly national concern. Documents from the federal Environmental Protection Agency (EPA) show that nationwide, 1.5 billion pounds of nitrogen from manure lagoons and another 880 million pounds of nitrogen from liquefied manure spread on land, ends up in surface waters after first evaporating into the air. Hog operations also emit 70,000 tons of hydrogen sulfide gas, 296,000 tons of methane, and 127,000 tons of carbon dioxide every year.

"Ya' know what really drives me nuts about the air pollution from hog factories?" Rick continued on his diatribe, "It's *intentional!*" "I don't get what you mean." Turns out, manure lagoons were actually *designed* with the purpose of getting rid of pollutants by releasing them into the air. Rick later showed me the industry's own documents that proved it. Dumping waste into the atmosphere, it seems, has long been part of agribusiness' plans. "And they're still not being forced to stop?" I asked incredulously after Rick explained the scenario to me. "That's right." "Good Lord."

In the days we drove around the state together, Rick provided me a running narrative about the many failures of liquefied manure systems that he'd witnessed in his years on patrol. "The state actually encouraged the lagoons, but that was just plain dumb," he commented. "In our wet climate, and with these sandy soils, lagoons don't work—they can't work. They're *doomed* to fail." The problem is this: Pigs poop a lot every day, so operators must regularly empty out their lagoons (by spraying or spreading lagoon contents onto land). However, when it's raining, it's impossible to spray the liquid or spread the slurry without causing pollution. "When the soils around here are wet, they're no better than a sieve," Rick continued. "Manure goes straight through to the groundwater and ends up in the rivers." This is why, even under North Carolina's permissive regulatory scheme, it's (theoretically, at least) illegal to spread manure when it rains.

The mound of studies I'd already compiled recorded risk after risk stemming from confinement animal operations and their waste. For one, as also shown by Rick's crews in North Carolina and Terry Spence and his team in Missouri, the waste contains high levels of the nutrients nitrogen, phosphorous, and potassium. *Nutrient* is a word I'd been accustomed to viewing in a strictly positive light. But in both Missouri and North Carolina, I was seeing how industrial animal operations cause far too many nutrients to enter groundwater and surface waters, which in turn was triggering episodes of hypoxia that were killing fish and other water animals.

My collection of studies also talked about manure from industrial animal operations containing other contaminants. It can have high levels of heavy metals, especially copper, along with residues of various drugs and chemicals. The primitive waste handling method used by confinement operations ensures that anything fed to hogs (including pharmaceuticals) or used in the confinement buildings (such as disinfectants) will end up in our soils and has a good chance of ending up in public waterways, groundwater, and the air.

Manure lagoons have also been shown to be ideal hatcheries for virulent viruses and bacteria. As I would learn more about later, almost all confinement animal operations continuously administer antibiotics in feed or water. The practice is done both to speed growth and stave off diseases. Therefore, the urine-feces slurry is laced with both antibiotics residues as well as pathogens that have evolved to survive these drugs. These infamous antibiotic-resistant "super-bugs" can cause treacherous diseases for both people and animals. Illnesses that were once easily treated with antibiotics are becoming debilitating or even fatal.

My studies also taught me that turning pig poop into a liquid was at the core of many environmental and health hazards caused by lagoons. A report by the Institute for Agriculture and Trade Policy noted: "The intentional mixing of water and animal wastes is proving to be one of the great mistakes of modern technology." The report goes on to explain that liquefied manure lowers industrial operations' costs but more readily transports nutrients and other contaminants to groundwater and surface water than manure in its solid form.

Pathogens are a particular concern. At traditional pig farms, the germs in pig manure are killed in composting or by sunlight when pigs are on pasture. This does not happen with lagoons, notes the report. "Unlike the composting or heating that takes place naturally in stored manure from animal housing when ample straw bedding is used, manure stored anaerobically in liquid form never reaches the temperatures necessary to kill pathogens [and] parasites." Lagoons, then, are hazardous both in the *quantity* of manure stored and in the *quality* in which it's stored.

GETTING A FISH'S PERSPECTIVE

After spending three days reviewing DNR documents with Rick, I was finally rewarded with some real adventure: accompanying the colonel on patrol. We would start on the water. Donning wind-breakers and Waterkeeper caps, Rick and I climbed into his mid-sized fishing boat and he fired up the engine. We slowly pulled away from his dock and were soon skimming over miles of dark, choppy river water, first heading a ways south, then north. It was a sunny September day with a crisp breeze.

At several spots, Rick cut back the motor and glided into inlets where he pointed out algal blooms and other telltale signs of pol-lution. All the while, like the proud father of an errant child, Rick boasted of the river's admirable traits while lamenting its prob-lems, most of which have been caused by recent decades of various human activities, industrial farming in particular.

I could see that Rick relished his moments on the water. The Neuse is North Carolina's third largest river, up to six miles wide and 250 miles long, and, even in its impaired state, hosts abundant wildlife. Having spent much of his time in his boat in recent years, Rick has seen it all: egrets, pelicans, herons, menhaden, shad, her-ring, catfish, bass, flounder, blue crabs, oysters, alligators, even dol-phins and an occasional manatee. As we rode around in his boat, Rick rattled off a battery of statistics about the river. He especially emphasized one fact. "The Neuse feeds into estuaries that are the most important fish nurseries in the Eastern United States. Just about every major commercial Atlantic fish species has sizable spawning stocks here." The significance of this was clear—a single river's con-tamination was compromising the health of national fisheries.

We did not have to go far to see the effects of pollution. Rick could simply reach over the side and pull out clumpy green hand-fuls of algae that were choking the river inlets and entangling the boat's propeller. "This is from excess nutrients," he explained, adding, "but it wasn't like this when I first came to North Carolina."

These days, a variety of sources put nitrogen and phosphorous into the Neuse River. The single largest—responsible for 56 percent of Neuse River nutrient contamination—is agriculture.

My river tour connected the dots between industrial animal facilities and our nations' waters and gave me a good overview of the resulting environmental damage. After a while, in the soft afternoon light, Rick reluctantly pointed the boat toward home. When we returned to the little pier behind his house, Rick pulled out a fishing net. Standing on the dock, he cast it into the water. "There's something else I want you to see," Rick said. He pulled up the net dripping with water and a dozen or so silvery fish, each about the size of a banana skin. These were menhaden, an important part of the Atlantic food web. About half of the flopping fish had gaping sores on their sides or backs. Once we'd examined them, Rick quickly threw the fish back into the water. But their blue and crimson innards were severely exposed, dooming them to certain, imminent death. "That's from *Pfiesteria*," said Rick, a heaviness in his voice for the first time. "Another result of excess nutrients in the water." For this one task, Rick had put on thick, black rubber gloves up to his elbows. It was *Pfiesteria* that had made him sick years earlier when he'd tried his hand at commercial fishing.

I had also read up on *Pfiesteria piscicida* before making the trip. The name roughly translates to "fish killer." The tiny organism (technically, a toxic dinoflagellate), had been discovered and christened by Joann Burkholder, a North Carolina State University biology professor whom Rick had befriended. She and her team documented how the organism liquefies fish flesh while giving off dangerous vapors. For fish, *Pfiesteria* is fatal; for humans, exposure to the emitted toxins can cause nausea, dizziness, and even neurological damage.

Dr. Burkholder had also documented a 500 percent increase in ammonia in the Neuse River estuary over the preceding decade. Her research related the rising ammonia and *Pfiesteria's* emergence to nutrient pollution from industrial animal operations. After she published her findings, the pork industry swiftly and sharply retaliated. Almost overnight Joann Burkholder's character and work became the object of vitriolic attacks (later written about in a book called *And the Waters Turned to Blood*). But agribusiness' attempts to vilify Dr. Burkholder could not change the facts on the ground: Massive numbers of fish were surfacing with sores and dying, and fishermen and lab workers exposed to *Pfiesteria* infested fish and waters were becoming terribly, perhaps irreversibly, ill.

After showing me the infected fish, Rick said he wanted to introduce me to some of the afflicted fishermen. The next morning, we drove to one of their homes, where Rick had arranged for them to gather. We sat down in the main room of a modest bungalow with three men in their mid- to late-thirties. Each had quarter- and nickel-sized sores on their arms and legs, some covered with Band-Aids. One by one the men told their stories about how they'd come down with the sores after consecutive days of fishing. Then they awkwardly pulled back their bandages, uncovering their dripping wounds. The sores looked startlingly similar to the ones I'd seen on the fish a day earlier. "They just don't heal up," one of the men mumbled with embarrassment.

The most unforgettable of the three was David Jones. His pale, frail limbs, splotched with dark sores, jutted out from his shirt and shorts. Even more worrisome was neurological damage. He said one evening, after several days of fishing in infected waters, he could not remember his way home. When I met David, months after that incident, he appeared to be far from recovered. As Rick and I spoke with him, his eyes wandered around the room and he seemed to have trouble focusing. Actually, he had become a totally different person, his wife told us in a low voice as she walked with us to the door. After thanking the fishermen and saying our goodbyes, Rick and I drove back to his house in silence.

INSIDE THE BATCAVE

That evening, we watched the news and ate dinner with Rick's wife, Joanne, a jovial, energetic woman who evidently supported Rick in countless ways. He and I then headed to his office, a separate building tucked behind their house. It felt like entering the Batcave. Everywhere were computers, monitors, video cameras, tripods, DVD players, telephones, and rows of storage cabinets and shelving filled with binders and videotapes. I half expected to see a red phone connecting directly to the police commissioner's office.

Rick began pulling down stuffed binders and placing them in front of me. "You should take a look at these," he urged. "It would take days to go through all of it. I've got hundreds of hours of video and tens of thousands of pictures." The binders were brimming with photos Rick had taken over the preceding years. "We're in a target-rich environment here," he commented in his military lingo as I began flipping through the pages. "Truth is, our work is only scratching the surface of all the illegal pollution going on around here every day."

I perused the pictures for almost two hours. Most had been taken from the vantage of boats and airplanes. Some showed huge cannons spewing a muddy liquid directly into creeks and streams. Others showed a massive brown plume of hog manure invading the Neuse River after a hurricane. The most disturbing were those showing scores of bloated pigs floating in filthy floodwaters.

"It's ridiculous that hog factories can even be built in flood plains, but that's exactly where they're put because that's where the land is cheap," Rick observed. "It's just asking for trouble. They're guaranteed to get flooded. It's just a matter of how often. And they always leak."

Leaking lagoons turn out to be a major way that hog operations contaminate groundwater and rivers. Studies from several states (including North Carolina) have demonstrated that manure lagoons continually leach pollutants to groundwater, tainted water that in turn enters rivers. In many cases, such leakage ends up contaminating someone's well, which may be their only source of drinking water. As I'd learned already, lagoons also pollute waters by releasing substances to the air that later end up in waters.

The third pathway to pollution is catastrophic spills. Whereas leaching and air emissions are the persistent daily drip, drip of lagoon pollution, spills are the periodic cataclysmic disasters. Every year, throughout the country, many lagoons burst or flood over. The consequences can be devastating.

North Carolina has surely seen its share of manure lagoon spills. In 1995, a lagoon broke open and spilled more than twenty-five million gallons of liquefied hog manure into the New River (more than twice the volume of oil spilled by the Exxon Valdez). According to university researchers who studied the spill, its noxious and foul-smelling plume traveled more than twenty-two miles and had noticeable effects for more than three months. The researchers also found that the spill killed millions of fish, triggered dangerous toxic algae blooms, and caused elevated fecal bacteria in the river waters and sediments. The infected waters put human health and aquatic ecosystems at risk while causing an estimated $4 million loss to the recreational fishing industry alone.

A couple of Rick's photo albums were dedicated to Hurricane Floyd pictures. When the hurricane hit the Atlantic coast in September 1999, Rick had been on duty as Neuse Riverkeeper for several years. As soon as the storm subsided, Rick began flying over flooded areas, taking photos and video of hog operations. Some of his footage even ended up on national news broadcasts. Rick was shocked by the devastation he witnessed. One of the most troubling things he saw was more than fifty confinement hog operations inundated by floodwaters. Their manure lagoons and flush systems were contributing hundreds of millions of gallons of liquefied feces to the waters that covered large sections of the state. Newscasts were reporting that these filthy flood waters posed an unprecedented threat to human health.

What struck me most about Rick's photo collection was the sheer volume of incriminating pictures. State and federal laws make dumping hog waste into waterways illegal, yet his pictures showed clearly that the laws were being widely flouted. And no one other than Rick Dove seemed to be doing anything about it. It appeared that Rick could go out on any day of his choosing and find operations flagrantly violating anti-pollution laws. I could see now that our legal cases would have a firm footing.

The hog industry, it should be noted, has steadfastly denied that its operations have any impact on water quality (or cause odor, or any other problem). In a typical move, in 2004, a group of North Carolina confinement hog operators funded a $30,000 study purporting to show that hog facilities were having "no effect" on the state's rivers. Notably, the analysis was not done by aquatic biologists or by anyone who'd ever engaged in any firsthand research on North Carolina rivers.

In response, the state DNR pointed out that for the four watersheds considered, $500 million of public money had already been spent to combat water contamination. Pollution problems were so severe in those areas that instead of seeing the marked improvement one would expect from a half-billion dollar investment, water quality in those areas was the same or worse. In other words, the industry-funded analysis was on its face absurd.

Such protestations of blamelessness prompted Dr. Bill Showers, a leading North Carolina water biologist, to comment, "It is going to take several decades before we realize the full impact of the hog industry on the environment." Dr. Shower's research traces industrial animal pollution using a nitrogen isotope found only in animal waste. With this technique, he has documented that nitrate levels from hog waste are as much as ten times higher downstream from hog operations than upstream. "To say that [industrialized hog operations] are not exporting nitrogen is hydrologically impossible," Showers has stated.

Dr. Mike Mallin, a professor at University of North Carolina's

Center for Marine Science, would certainly agree. For a decade, he regularly checked water quality at dozens of sampling stations on rivers in eastern North Carolina, many of which flow past scores of hog operations but little else. During that time period, Mallin and his team found a 265 percent increase in ammonia in the Black River and, in the nearby Northeast Cape Fear, an ammonia increase of 315 percent, river basins which have some 5 million confinement hogs.

A BIRD'S-EYE VIEW

After almost a week working alongside Rick, my last full day in North Carolina had arrived. Finally, it was my chance to be part of his famous air force. It would be my first time in a small plane and I mentioned on the way to the airport that I hoped we had a good pilot. "Don't worry," Rick said, "Ron's taking us up today. He served two tours as a pilot in Vietnam." I wasn't sure if that was comforting, but at least, I figured, the guy was unlikely to panic in a crisis.

Ron was a lively man with thick forearms and ruddy cheeks who wore a Hawaiian shirt and a backward baseball cap. He and Rick fell easily into conversation and mutual ribbing. Out on the tarmac, Ron busied himself with final checks of his aircraft like a cowboy tightening the cinch on his horse's saddle. His confidence and obvious familiarity with the Cessna assured me we were in good hands. Ron's enthusiasm about our mission, I sensed, came partly from allegiance to the colonel.

Loaded down like a burro with high tech cameras and telephoto lenses, Rick clambered into the backseat, as he instructed me to sit up front for the best view. I slid into the copilot's seat, buckled up and pulled on a set of headphones with a tiny microphone at the mouth. As we checked the mics, it felt like we were about to try a high tech simulator. But this was entirely real, and soon we were in the air.

Prior to takeoff, Rick had instructed Ron where to fly. Several destinations were the operations Rick regularly monitored and photographed from the air; others, he had recently heard were illegally polluting. As we approached our first target, the sun glinted off the roofs of two big metal sheds and a large manure lagoon. A short distance away an enormous water cannon was lavishly spewing a thick brown fluid.

Rick strained toward the window exclaiming, "Aha! Just as I suspected." He raised his camera and asked Ron to go in low. Ron complied by tipping the plane's nose downward, and we plummeted toward the earth. Our pilot's wartime experience seemed to be coming in handy. This was nothing like flying United, and I couldn't help grabbing the sides of my seat as a surge of warmth rushed through me.

As we hurled toward the hog operation beneath us, a blue pickup suddenly came into view. The truck was racing toward the manure sprayer, kicking up a billowing trail of dust in its wake. We were now at what seemed about 500 feet and Rick's camera was clicking away madly behind me. "Circle back!" he yelled, and Ron complied by banking the plane to the left in a wide arc. My heart was racing.

On the second pass, we seemed to plunge down even lower. But by the time we made our second overflight, the sprayer had been shut down. A man standing near the spray gun was staring up at us with his hands on his hips. He had jumped out of the pickup and run over to turn it off. "He knows exactly who we are!" I blurted with pride. "You bet he does," answered Rick. "This feels like warfare!" I exclaimed. To which Ron soberly replied, "It is."

Forty minutes later, after reconnaissance on a few more operations, Ron glanced over at me and nonchalantly said, "Now it's your turn." I smiled, assuming he was joking. But a moment later he took his hands off the U-shaped steering control, leaned back into his seat, crossed his arms on his chest, and added, "You're in charge now." I gulped and took hold of the controls in front of me as I felt my palms getting clammy. Within seconds, we were rapidly losing altitude. "Steady her," Ron advised calmly. "Keep your eyes on the horizon." With those wise words, my competence in keeping the craft level quickly improved.

We flew over a few more hog facilities (with Ron back on the controls, of course) while Rick snapped photos. As we started toward home, Rick suggested we try to count the manure lagoons we could see from this bird's-eye view. The land below us was littered with them. Each of us lost count around one hundred.

Back on the ground, I was walking on air. "I got some good shots," Rick informed us matter-of-factly. "Those guys were spraying manure right into a stream." For Ron and Rick, this was all in a day's work, but I was thrilled that we had accomplished our mission and returned safely to base.

ITCHING FOR A FIGHT

On my trip back to New York, I realized there was nothing I'd rather do than pick a fight—a really big one—with the animal confinement industry. My initial suspicion that I'd be drowning in manure was partly right. Industrial operations warehouse throngs of animals that generate oceans of liquefied manure. A University of North Carolina professor has calculated that an adult hog produces as much as ten times the waste of an adult human and that the largest hog complexes can actually generate more animal waste every day than the human feces produced in New York City or Los Angeles. But cities collect and treat their sewage according to strict guidelines. Hog operations do not. They discharge their manure into the environment essentially untreated, a difference unwarranted by science. So, manure would play a leading role in my future.

But it was becoming clear to me that this was not a battle against animal waste. After all, manure is not inherently loathsome. It's all a matter of how much of it is in one place and how it's handled. I'd once heard a doctor say about aspirin, "Toxicity is all about dose." The same can be said of manure.

And I now saw that avoiding meat didn't make me unconnected to this national disaster. I had ties to the confinement food animal industry because I still ate eggs and dairy products. Even more importantly, I realized that these were issues close to my heart. Battling factory farms is a fight for people wanting to enjoy the sanctity of their family homes, for protection of lands, waters, and air, for citizens' rights to govern themselves, and, perhaps most of all, for billions of farm animals who never know a moment of joy.

I was now eager to engage this industry, which seemed to be rampaging out of control and unchecked. As I drove to my apartment from La Guardia that evening, I called Bobby from my cell phone with my answer. "I'll have those pleadings for you in a week."

CHAPTER TWO

An Unexpectedly
Natural Turn

MY UNREALIZED ROOTS IN
FOOD AND FARMING

The events that led me to the job in New York had started one
drizzly winter evening back in Michigan in early 1999. That night
I hurried into the local university's student center where I'd been
invited, as a city councilor, to sit in the first row for a talk by Robert
F. Kennedy, Jr. A member of such a well-known family generated
excitement in Kalamazoo, and there was a palpable buzz in the air
as I made my way to the front of the auditorium. I had surmised
that the speaker was the son of the late Senator Robert Kennedy,
but it was the first I'd heard of him, so I wasn't sure what to expect.

As he stepped to the podium Kennedy's physical resemblance
to his father was almost jarring. But then he uttered a few words.
His voice cracked and he paused, cleared his throat and began
again. My heart sank. This Kennedy seemed to lack his family's

gift for inspiring oratory. My doubts, however, would be quickly and permanently dispelled. I soon found myself leaning forward in rapt attention as he unfurled tales of his adventures saving us all from unscrupulous corporations that were, as he said, "treating the planet as though it were a business in liquidation." It gradually dawned on me that this man was describing just what *I* should be doing.

Hearing that speech altered the course of my life. Within months, I went to work as an environmental lawyer. Within a year and a half, I'd moved far from where I'd settled to begin working as the senior attorney for Kennedy's organization. And shortly there-after, my *raison d'etre* became remaking the modern food system.

Looking back at it, though, these unexpected turns seem like smooth bends in my path. My parents instilled in each of their four children an abiding respect for nature and a passion for food. "Mutti" (as we call my mother, who's from Germany) has always spent many hours tending her gardens. With a little help from us kids, she grew a fair amount of what she fed us: tomatoes, lettuces, cucumbers, green beans, chives, squashes, and the occasional (usu-ally unplanned) pumpkin. Our fruit she insisted we pick at local orchards. From June to October, while other parents were taking their kids to amusement parks, mine were running a mini-migrant labor camp. We started with strawberries, then moved smartly into cherries, blueberries, plums and peaches, and finally ended the season with pears and apples. Bananas, oranges, pineapples, and other very non-Michigan produce were strictly relegated to winter dining. Eggs and those vegetables she didn't grow herself, Mutti bought directly from farmers at farm stands.

We spent time on other local farms, too, like that of the Moores, a young couple and their son with whom we'd become friends. The Moores were struggling to keep their farm afloat. The Hahn clan would squeeze into my parent's Oldsmobile sedan (considered plenty large for a family of six in those days) to head west of town for a day's labor at the Moore farm. My parents hoped to help them out while giving us a true taste of farm life. I fondly remember riding a paint pony, gathering eggs, and digging potatoes from the black earth. Some of the experiences were less pleasant. I was heartily zapped by the Moore's electric fence, something a child does not quickly forget.

From our time on farms, my siblings and I learned what came in season when. No one in our house was under the misimpression that food originated from a Styrofoam tray or a cardboard box. Mutti canned much of our harvest, baked her own breads and pies, made yogurt, and picked and dried herbs for cooking and tea. Nature's role in our food was always quite clear.

Eating together was a daily ritual. Every evening, my brother, two sisters, parents and I would gather round my mother's oval table for a home-cooked meal. It was at the dinner table that our mother and father peppered us with annoying parental questions like, "So, did you learn anything at school today?" and where most of the sharing of information, spilling of confessions, and tattling took place. In our teenage years it was where my sister ratted me out for having gone to a forbidden party and where I retaliated by telling on her for having run a red light in my parents' powder blue Ford Grenada. It was also where, almost two decades later, I let my parents know that their long wait was over—I'd finally met the man I wanted to marry. The dinner table has always been our family hub.

My father's connection with nature, while less pragmatic than my mother's, is equally intense. He adores all types of natural landscapes, whether cultivated or untouched by humans. It's a deep-seated passion he developed as he whiled away his youth in the company of his dog Bozo, exploring streams and roaming rolling woodlands northeast of Cincinnati. Our annual family pilgrimage to my father's homeland in Ohio was invariably made entirely on small country roads instead of freeways, although it almost doubled the length of the drive. This was much to the dismay of his pack of children. "Just look out the windows!" he urged with exasperation every time one of us whined.

For forty-some years, whether sunshine, rain, or snow, "Vati" (as we call my father) and his dogs took long daily walks together in a field and woods near our house. Whenever I accompanied him, which was hundreds of times over the years, Vati paused to pay equal homage to nature's events, tiny and grand, along the path: an insect crossing, a bird song, or the warm golden glow of the twilight sun. Mid-stride, he would freeze, lower his voice, and point with a squint, "Look! See how the sun's beams are striking that pine!"

Vati often waxed admiringly on Indian trackers who pursued animals for miles based solely on a broken twig here, some crushed grass there. He always said his favorite childhood book was James Fenimore Cooper's *Last of the Mohicans*. Vati wanted to teach us to be like those original Americans—appreciating all living creatures and noticing every sight, sound, and smell surrounding us in the wild.

My father's beloved patch of nature was my favorite childhood classroom and church. I spent hours there in his company as well as on my own. Part of the land had once been farmed and Vati enjoyed showing me the location of every old raspberry bush, choke cherry, apple, and pear tree. I sought out their ripe fruits in the season like hidden jewels. Year after year as I wandered those forests, lay in the meadows, and plucked wildflowers from the fields, I observed nature's cycles: seedlings sprouting, tree lines marching forward, old oaks dying, dead logs decaying. One day I confided to Mutti that I felt most certain there was a God when I walked those woods.

In my adult years, an indelible tattoo of this seemingly unremarkable two hundred acres revealed itself. Ever since leaving home for college, I've often found myself walking or playing there in my dreams. When wide awake, those forests and fields have occasionally appeared before me in flashes.

Vati's de facto nature preserve laid claim on my soul. It literally drew me back from North Carolina a couple of years after law school. While a community battle raged over a plan to clear the land and erect flimsy student apartments there, I returned with the idea of stepping into the fray. I felt compelled to try to protect a place that had become almost sacred to my father and our family. When I got back to town, I started holding meetings with neighbors, and eventually formed a citizens' group to preserve the land.

One woman in our little association had a long history as a fly in the ointment on local issues. She thought that to get our concerns addressed I should try to get elected to the City Commission. The suggestion sounded a bit outlandish since I'd never run for office and had been back in town only a few months. But with a truly grassroots campaign, we did manage to get me elected. As a city commissioner, I helped craft a plan to protect much of the land I'd been trying to save. My father, now in his mid-eighties, walked there every day until the cruel effects of advancing age robbed him of this pleasure.

By the evening I first heard Bobby Kennedy speak in Kalamazoo, I'd been back in town about four years, and was groping around for my next step. I worked at a respected law firm and devoted much of the rest of my time to city council business. Between that, spending time with family, being a homeowner, and competing in triathlons, my life was contented and certainly crowded with activity. But it still didn't feel quite fulfilled. I longed to put more of my energies toward things that I believed would really matter over time.

Representatives from both the Republican and Democratic parties had approached me about running for higher office. But the morsel I'd tasted of an elected official's life gave me the definite impression that it was nothing to lunge at. Endless hours of work, dubious forward motion, and very little thanks at the end of the day. Higher office sounded a bit gloomy, particularly for a thirty-something single woman living in a Midwestern town.

I had majored in biology in college and had long been active in environmental causes, especially on the city council. But my job, as a business lawyer, was far removed from my passion for nature. Bobby Kennedy's tales of attorneys fearlessly battling to stop pollution and protect wildlife cast a beam of light on my path. I felt called to join the cause.

Shortly after hearing his speech, I sent Bobby my resume and a letter that closed by stating, "I would like to discuss the possibility of working with you." About two weeks later, he called me at my law offices. "Hello? This is Bobby Kennedy." I could tell that it actually was by the unmistakable gravely voice. "I really liked your letter. You have just the sort of background I look for," Bobby said. "I don't have anything for you right now but I'm going to keep your resume in a very thin file." *A very thin file!* The call lasted no more than three minutes, but I was elated. Writing Kennedy had seemed a shot in the dark, especially since I had no experience in environmental law.

That phone call infused me with confidence to follow my calling. Three months later, I took a job as a lawyer at the National Wildlife Federation (NWF) and shortly thereafter moved to Ann Arbor. I became part of their team that focuses on protecting the Great Lakes, using, in particular, the tools of the federal Clean Water Act and Clean Air Act.

Accepting the job with NWF had meant leaving my law firm, retiring from the city council (at the close of my second term), selling my house, and moving from my hometown where I had a strong web of friends and supporters, as well as my parents. But the decision felt absolutely right. I loved the work and soaked up the new knowledge and sense of purpose.

Eleven months later, a group of my NWF coworkers and I went to hear Bobby Kennedy speak at a local college. "This'll be like attending a good sermon," I prepared my colleagues. After his talk, I introduced myself, reminding him of our phone conversation a year earlier. To my astonishment, he remembered. To my even greater astonishment he said, "Hey, I've got something for you now. I can get you in on the ground floor of a national organization I'm putting together." He was talking about Waterkeeper. It sounded strangely like the pitch of someone trying to lure me into a pyramid scheme, but, admittedly, I was intrigued.

Ten days later, I flew east for a weekend to meet the people I would be working with if I accepted the Waterkeeper job. It seemed like a vibrant, dedicated group.

Still, I hesitated because I felt committed to National Wildlife Federation. I decided to confide in a kindhearted colleague, Tony. As the main partner on my project, he would be the most affected by my departure. Tony's advice was unequivocal. "*What!?* You're even *considering* not taking this!? This is your dream job. *Just go!*" He made it all so simple. I said, "You know, you're right, I think I will." And I did. Two months later, I was off to New York.

CHAPTER THREE

What Came First, the Chicken . . . and the Egg

POULTRY ENTREPRENEURS PIONEER FACTORY FARMING

Within a few weeks of being tapped to start a campaign against industrial animal production, it was consuming my life. Creeping into my meals were questions about the origins of every animal product I ate. Visions of unappetizing pig operations danced in my head. Photos and videos I was looking at every day on the job were making clear that other types of animal operations, too, including egg-laying facilities and dairies, were equally bleak places. It had become impossible to ignore the shady past of my own food.

But I wasn't sure what to do about it. I was not ready to give up staples like omelets, yogurt, and ice cream. (Not to mention butter and cheese.) And, more fundamentally, I had never embraced the argument that animal farming was inherently wrong. It was *this in-dustrial way* of raising animals I found so troubling.

At the same time, the demands of getting the project under-way were stretching my work day well into the evening. When I would finally make it back to my apartment, usually long after dark, I often lay in bed devouring gory tales of the meat industry. A pen and notebook were always ready at hand for underlining and note-taking.

As I began constructing our public and legal cases against in-dustrial animal production, I kept wondering how the business had gotten so warped. Without exception, every time I told friends and family about what I'd witnessed in Missouri and North Carolina, they were shocked. This led me to ask myself: How is it possible that our food is being produced in ways so out of sync with our values, and with none of us realizing it? As a history professor's daughter, I knew there was a story here and decided to investigate it. I quickly discovered that the shift toward industrialized animal production started with poultry farming.

My first-hand experiences with chicken and turkey farming had been scant. When I was five, my Aunt Margaret (rather unwisely) brought my sisters and me an Easter gift of a small paper bag con-taining two fuzzy yellow peeping chicks. Regrettably, but not sur-prisingly, both died within a few days. The experience was not very instructive for learning the ways of fowl husbandry. Around age ten, my family made an outing to a local turkey farm. It was a place that welcomed visitors and even had a cafeteria that served turkey sandwiches. The turkeys freely roamed a large fenced yard. I re-member being struck by their grandeur and friendliness.

On the Moore farm, I had wandered amongst red, white, and black chickens in the barnyard and observed them in action. They pecked and scratched the ground searching for bugs and grain, stretched their wings, preened their feathers, and doused them-selves in dirt. (My father had explained that, paradoxically, this was part of their cleaning ritual called "dust bathing.") I loved collect-ing their warm eggs, which arrived in an array of sizes and pastel shades.

From these few encounters with chickens and turkeys, I knew little about these creatures and how they were raised. But gaining some insight into poultry farming was critical to understanding animal farming's industrialization. When Waterkeeper held a press conference to kick off our reform campaign, the National Farmers Union president joined us, proclaiming he would not stand idly by while the hog sector underwent "chickenization." For him and others (supporters and critics alike), poultry production had become a model, even a mascot, for mechanizing animal farming. I would soon learn that the poultry sector's industrialization was a gradual process, starting near the tail end of the Industrial Revolution. My research took me back even a bit earlier.

CHICKEN FEVER

Few Americans today have firsthand encounters with live chickens, but that was not always the case. Far from it, in fact. In the middle of the nineteenth century, chickens were all the rage in the United States, and not just among farmers. To make what I'm about to say believable, one must keep in mind that these were times when people had limited entertainment options—no movies, no television, and no computers. Chickens, it seems, were to the mid-1800s what the Beatles were to the 1960s. That may be a slight exaggeration, but you get the idea.

At that time, small chicken flocks were kept behind many houses, even in cities. The birds and their eggs appeared in a rainbow of colors, shapes, and sizes, with as many as five hundred breeds in existence, according to the American Livestock Breeds Conservancy. Typically, women and children tended the flocks. They fed the chickens vegetable peelings, table scraps, and a few handfuls of grain, and made sure the birds were securely in their coops at night. This method of care and feeding made the costs of keeping chickens negligible. The work was rewarded daily with eggs and occasionally with meat. Contact with chickens was about as commonplace as interactions are today with dogs and cats.

The Boston Poultry Show of 1849, however, where more than a thousand varied birds were on display, whipped this familiarity into a veritable chicken mania. In the decades that followed, "poultry exhibitions were held in every city, state and county, and ribbons so bitterly contended for that hardly an exhibition escaped charges of bribery and favoritism," according to one chicken history narrative. By the late nineteenth century, Cornell, Penn State, University of Connecticut, and numerous other major universities were offering classes in poultry husbandry. By the early twentieth century, 65 colleges and U.S. Department of Agriculture experimental stations were researching poultry husbandry. And between 1870 and 1926, more than 350 different periodical publications were introduced that dealt solely with the raising and breeding of poultry.

THE ERA OF FARM CHICKENS

Inevitably, this chicken fever cooled. But poultry populations continued to swell and chickens remained virtually omnipresent occupants of rural and urban America well into the twentieth century. An 1880 census, counting poultry for the first time, recorded 102 million birds. The number popped up to 258 million by 1890, then climbed more gradually. By 1910, a farm census showed about 280 million chickens and a value of poultry products exceeding $500 million, an annual revenue second only to corn among agricultural products.

At the dawn of the twentieth century, then, chicken ranks were rising and they were widely dispersed. A 1906 census showed that in urban areas there was one chicken for every two people, while the 1910 survey noted that 88 percent of farms kept chickens, with flocks averaging around 80 birds. This first part of the twentieth century has been described as "the great era of the *farm chicken*," as opposed to the era of *chicken farms,* which would come later. These were the days when farms actually looked a lot like the image many of us hold in our minds of where our chicken and eggs come from. One poultry industry chronicler writes that 1900 to 1940 was "the tranquil period in American life when almost every farm had a small flock of chickens to provide those legendary farm breakfasts, with a goodly number of eggs left over [for market] to provide the necessary spending money ('egg money') for the woman in the family."

AUTOMATION AND SPECIALIZATION

In spite of its tranquility, in this period's history I also detected the first dark hints that poultry farming, then eventually all animal farming, would be industrialized. For one, certain voices (usually those who stood to gain financially) urged that particular poultry breeds should be raised for eggs while others are raised for meat. The concept would be unworkable—silly, even—in a traditional farm setting where each chicken served multiple purposes for the farm. But specialization fit in the newly forming industrial approach with its focus on efficiently producing as much as possible of one thing at the lowest cost. What Henry Ford was doing with his Model-T, in other words, was being applied to agriculture.

One pillar of industrial production was specialization, another was automation. For farming, this meant taking activities that had once been carried out by humans or the animals themselves and instead doing them with machines. Automated feeding and watering systems are one example. The most successful innovators in this arena were the DeWitt brothers in Zeeland, Michigan, who invented automated systems known as Big Dutchman feeders. Where such automation was adopted, the chickens required and received much less human care, which the DeWitt brothers and others touted as a great advance, wonderful human progress.

These were physical indicators of industrialization. But equally important, perhaps more so, is the accompanying attitudinal shift that I detected emerging at this time. When families lived near and cared for a small flock, they almost couldn't help coming to know the individual birds, noting one chicken's friendliness, another's churlishness. The numerous children's stories with principal poultry characters reveal this recognition. "The Little Red Hen" suggests a hen's legendary diligence, and "Henny Penny" (also called "Chicken Little") depicts some chickens' skittishness. People noticed these character tendencies because they lived in proximity to their birds and because it behooved them to do so: They always wanted their next generation of breeding animals to be the offspring of their best hens.

And as long as the chicken was still considered as an individual, humans also could not avoid noticing that he or she had certain needs. Consider, for instance, the 1867 *Practical Poultry Keeper*, which opens with the following advice: "Fowls should not be kept unless proper and regular attention can be given to them; and we would strongly urge that this needful attention should be *personal*." The guidebook continues by regaling its readers with pearls of wisdom about the best ways to care for domesticated birds, including a detailed description of how to prepare a hen's nest. If not done just so, the *Practical Poultry Keeper* warns, "the hens will often drop their eggs . . . upon the ground rather than resort to them."

However, as flocks expanded from dozens to thousands, and interactions between fowl and humans dwindled, the very idea of the individual chicken seems to have evaporated from the farming lexicon and mindset. Advertisements, articles, and books of the day tout strategies for increasing the pounds of meat or eggs produced. Increasingly, the focus was on the amount of feed used (feed conversion). Two calculations measured success in the newly emerging commercial poultry business: first, how many pounds of feed it took to produce a pound of chicken flesh or eggs, and second, how much it cost in total to raise a live chicken. Breeding of meat-chickens targeted rapid growth and high feed conversion rates, while breeding of egg-laying hens focused on getting more eggs using less feed. The art of chicken husbandry was being diminished to a math equation and a balance sheet.

As I dug through historical books and articles, it was apparent that such changes were the markers of traditional farming's transformation into an industrial enterprise. But I felt they failed to fully explain how humans had made the dramatic shift from an approach guided by nature's principles to an industrial model, in which nature seemed entirely banished.

Then one day I hit upon the key to chicken farming's original break with tradition: *the incubator*! Early Chinese, Italians, and Egyptians (among others) had done some egg incubation—that is true. But by and large, in the first two centuries of our nation's history, it was mother hens that hatched and raised chicks. The success of the mechanical incubator, which had been invented even before the automated feeder, depended on acceptance of the idea that the very essence of animal husbandry—the rearing of young—could be done without natural mothers or even human caretakers. In other words, it could be done entirely with machines. Around 1880, replacing mother hens with incubating machines began in earnest in several places, most notably Petaluma, California. As one history of poultry agribusiness says (approvingly), the incubator set poultry farming "on an unerring industrial course." The more I learned the more convinced I became that the incubator was truly a radical departure from traditional animal husbandry and its first definitive rupture with nature.

Chickens, of course, have been around for eons. Their ancestors are believed to have evolved tens of millions of years ago. Humans first domesticated them from the Red Jungle Fowl (scientific name, *Gallus gallus*) in Thailand or Vietnam some 10,000 years ago. Millions of years of natural selection, then thousands of years of human-guided selective breeding, had fostered the mother hen's intense desire to nurture her young. However, the incubator abruptly discarded this long distinguished history of motherhood.

In traditional chicken tending, the best hens were carefully selected and allowed to keep a clutch of eggs, on which they'd set for twenty-one days, until hatching. Hens then became famously fierce and loving caretakers of their chicks. "Broodiness" (mothering traits) had been valued by farmers and lauded by poets and philosophers through the ages. Consider Plutarch, who often wrote admiringly about chickens, observing that "there is no part of [hens'] bodies with which they do not wish to cherish their chicks if they can." And listen to the raptures of Oppian, a poet of ancient Greece:

> *With how much love the playful hen nourishes her tender young ones! If she sees a hawk descending, cackling in a loud voice, her feathers raised high, her neck curved back, [she] spreads her swelling wings over the clucking chicks. Then the frightened chick chirps and hides himself under these high walls, and the fearful mother gathers the long line of young chicks under her plumage; careful mother that she is, she attacks the bold attacker and frees her dear chicks from the mouth of the rapacious bird . . .*

The incubator, however, not only made mothering traits unnecessary, it made them undesirable. Farmers actually began to select mothering tendencies *out* of flocks. It was at this point in farming's history, I believe, that the animal stopped being regarded as part of a natural order and began to be treated more as a production unit.

The incubator, it dawned on me, likewise ushered in animal farming's era of segmentation and disconnectedness. Families who raised farm chickens had taken great pride in their birds and in the animals' bloodlines, which their owners were striving to improve over generations. A farm chicken's life would likely be passed in the care of one human family, on a single farm, and with the same fowl peers. But once the incubator came into vogue, a bird's life became segmented and utterly unconnected to any one place or any one set of humans. Eggs were laid by hens in one location then moved to large hatcheries where they were incubated in machines resembling giant ovens alongside tens of thousands of other eggs of mixed origins. From hatcheries, baby chicks were boxed and shipped like biscuits, by the hundreds, to far-flung locations around the country. No chick would ever know its siblings or mother, let alone be taken care of by her. Chickens had become as much a fungible commodity as widgets.

By the start of the 1930s, chicken farming was indisputably entering the era of mass production. A single incubating machine could hatch 52,000 chicks at once, and flock sizes for meat-chickens had swelled into the tens of thousands. Moreover, demonstrating clearly that the individual chicken's life was no longer valued, starting in this decade, hatcheries began to check the sex of baby chicks as soon as they emerged from their shells. For the egg-laying breeds, males were considered worthless; in many locations, they were simply thrown away. Around this time, the Petaluma hatcheries, for example, began drowning two million male day-old chicks every year.

I was so shocked when I read that, that I stopped and re-read it. But as I thought about it I realized that a highly specialized animal industry is inherently wasteful. That's because once breeds are designated "meat breeds" and "egg breeds" (or "meat breeds" and "dairy breeds," for cattle, as I would learn later), males of the non-meat breeds are automatically characterized as an unwanted industry by-product. I was baffled that anyone could consider a system "progress" that *throws away* millions of newborn animals every year—one of every two live births.

A NEW USE FOR MEDICINE'S MIRACLES

Along with the incubator, several other "miracles of science," particularly drugs and manufactured vitamins, were enabling novel factory methods for raising chickens. Common sense and countless generations of practical experience had long dictated that living creatures could not be kept healthy without exercise, fresh air, and sunshine. Disease outbreaks and other health problems (such as rickets, from sun deprivation) had always made it impractical to keep large, continually confined flocks. When farmers first began raising larger flocks in the 1920s, typical death losses quadrupled, jumping from 5 percent to 20 percent. "With commercialization and greater intensification have come [new] disease problems that ... cause more acute financial losses due to the numbers of birds involved," two professors of poultry husbandry wrote in the 1930s. "Chances of disease spread are materially enhanced." Nonetheless, the crowded housing method pressed on, with manufactured vitamins and pharmaceuticals substituted for healthy living conditions. All chicken raising moved in the direction of large, continually confined flocks.

One individual especially stands out for his vigorous advocacy in favor of chicken drug dosing. Joseph Salsbury was an enterprising veterinarian, originally from England, who landed in Charles City, Iowa. In 1926, he set up a mail-order business for a plethora of poultry medications. Drugs were barely used in chicken husbandry at the time, but Salsbury enlisted an impressive sales force to blanket the Midwest persuading farmers that they'd benefit from the remedies of "Old Doc Salsbury." And there was more. He wrote prolifically for farm journals and soon set up a popular "poultry course" that, by the mid-1950s, had been attended by more than ten thousand people in the business. The printed course guide came to be considered a standard health manual for the nation's poultry industry. It is hard to overstate Salsbury's influence.

In everything he did, Salsbury encouraged liberal use of medications. He and his sales force were among the first to urge adding drugs to chickens' daily feed. All of the pharmaceuticals he recommended, of course, he also peddled as products. One of these, Ren-O-Sal, would eventually (beginning around 1950) become the first drug added to animals' daily feed to prevent disease outbreaks and to speed growth. Old Doc Salsbury's drug sales ended up making him a wealthy man.

By the late 1930s, chicken raising facilities functioned quite like factories. Meat and egg production were now mostly specialized. Flocks raised for meat had tens of thousands of birds and were often kept confined round-the-clock. The U.S. Department of Agriculture, along with trade associations and journals, promoted the need for absolute uniformity in the meat and eggs. Birds were inoculated with a variety of vaccines to ward off the many diseases that plagued them due to crowded conditions.

Chicken feeding, too, had changed. Poultry were provided premixed mash that included ingredients like excess fish and slaughterhouse wastes. Drugs were not yet added to daily feed, but were regularly administered to whole flocks showing signs of illness. Not surprisingly, operators soon discovered that confined, mash-eating chickens were collectively losing their appetites. Around this time, therefore, it became common practice to add arsenic to feed. Arsenic, of course, is a poison, yet in low-level doses it acts as a powerful appetite stimulant. With factory-style feeding and housing systems in place, only the total dominance of agribusiness was required for the industrialization of poultry production to be complete.

PATRIOTIC PRODUCTION

At this point in the story my research needed to diverge a bit from poultry farming to world history. The United States was perched on the brink of entering World War II. Even before the attack on Pearl Harbor, President Roosevelt had begun using his famous fireside chats to motivate American farmers and ranchers to help the British with food (and fiber). "Food will win the war and write the peace," became a U.S. Department of Agriculture slogan. Feed and transport companies and hatcheries seized the opportunity and began running ads encouraging farmers to help the cause by growing "defense chicks" for food to be shipped abroad. Producing as much chicken as possible was now being characterized as a patriotic act.

Once U.S. forces were actually in combat, the connection between maximizing production and duty to country was solidified. "Those who could contribute to the war effort by producing chickens were expected to do so. It was their duty," writes a poultry industry chronicler.

Whether stirred by patriotism or the chance to make a buck (or, most likely, both) especially rapid expansion occurred in an area of Delaware, Maryland, and Virginia (known as "Delmarva") near Chesapeake Bay. Then and there, farm chickens were definitively displaced by industrial chicken farming. Delmarva's poultry farmers, using crowded, automated confinement buildings, became the primary chicken supplier to the Allied Forces.

Meanwhile, the federal government vigorously pushed increased output of eggs and meat chickens in every locale. "[I]n every county across America, [USDA's] agricultural agents ... were marshaled to promote greater food production. One of their primary targets was an increase in poultry production."

A contemporaneous account, by *New York Herald Tribune* food editor Clementine Paddleford provides an illuminating report—glowing, actually—on the poultry sector. Her 1943 article even enthusiastically likens raising chickens to producing weapons of war:

> *Today, the frying chicken is turned out in mass production with all the precision and regularity of army tanks and planes. Chicks are machine hatched—and by the millions—to take their place on the assembly lines of the broiler meat factories. ... By mass production methods the poultry manufacturers work to meet 1943 goals demanding five billion eggs and four-billion pounds of chicken. Last year's egg and chicken crop was the highest on record. This year the little red hen promises to do better—18 percent better, the government requests.*

As Paddleford's article suggests, American farm polices fostered production during wartime, and then beyond. Prior to the Great

Depression of the 1930s, the United States had promoted farming with laws like the 1862 Homestead Act. Farmers had been encouraged to clear forests, drain wetlands, and plow prairies. But the government had never directly subsidized farm production. And farmers, who were famously protective of their independence, liked it that way. Most would probably rather have thrown themselves into a pit of vipers than receive any direct government support. However, the desperate years of the Great Depression changed all that. Between 1929 and 1932, prices received by farmers fell by 56 percent. Roosevelt enacted relief for struggling farmers that included both crop support payments and production controls.

World War II then trumped these policies. As we've seen, maximum production was encouraged (all controls except tobacco were lifted by 1944), and crop values generally far exceeded the support prices. When the war was over, the Truman administration maintained price supports for specified "basic commodities" (including wheat and corn), but, under heavy political pressure, failed to reinstate production controls. And so began American agriculture's publicly funded overproduction.

VERTICAL INTEGRATION
ENTERS ANIMAL FARMING

Facilitated by these federal farm policies, which made grain cheap and abundant, increasing both production and "efficiency" remained the poultry industry's overriding preoccupations after the war. Its poster child was a Georgia man named Jesse Dixon Jewell. Industry insiders consider him the founding father of truly vertically integrated chicken production. Jewell thereby led the final phase of industrialization of poultry farming.

As a young man in the 1930s, Jewell's family animal feed company was faltering. To revive the business, he decided to try to develop a dedicated customer base. Jewell hit upon the idea of promoting chicken growing to desperate cotton farmers, whose crop value was plummeting. To accomplish this he used a network of salesmen who sold both feed and the idea that farmers should become "contract growers" (by this time chicks were no longer "raised," they were "grown"). Jewell would supply the birds and feed, while farmers under contract would supply the land, buildings, and labor. Farmers were also granted the privilege of owning the chicken manure (called "litter").

Jewell's enterprise was highly successful, so he took it a step further. As his sales force developed an extensive farmer network to raise the chickens, Jewell built his own hatchery and processing plant around 1940. That, he realized, would give him control over every aspect of production—from baby chicks to processed meat. He boosted his bottom line by recycling his slaughterhouse wastes as ingredients in his chicken feed.

By the mid-1950s, J.D. Jewell, Inc., was famous for vertically integrated meat production. "Jewell controlled every phase of his operation from laying flocks, through incubation, raising the chickens, warehousing and distribution, sales and advertising, by-products, field crops, feed mill, specialties, and the experimental farms." He also became widely known as an energetic promoter of the industry who proudly invited the world to come see his company's operations.

Jewell's renowned openness, combined with his active involvement in regional and national trade associations, ensured that everyone in the feed industry was well aware of what he was doing. Others soon flattered him by imitation. All the while, U.S. Department of Agriculture extension agents actively supported entrepreneurs' efforts at vertical integration. With such encouragement, other Southern poultry operations soon followed Jewell's lead. They "laid the foundation for an agribusiness concept—the idea where a farmer-businessman team would produce food efficiently and effectively on an assembly line basis," writes an industry insider. Other regions were not far behind. With full vertical integration as the dominant business model, poultry's industrialization was now truly complete.

Largely due to Jewell's efforts, from 1950 to the mid-1960s Georgia was the South's leading meat-chicken state. North Carolina, Alabama, and Mississippi were following suit, developing chicken industries based on vertically integrated agribusiness companies and farmers raising chickens under contract. In each state, the industry's growth mimicked the pattern established by Jewell: Feed businesses promoted meat-chicken production and coordinated their marketing with regional slaughter plants. Turkey production, although a much smaller industry, was going down a parallel path.

The egg business was likewise copying the meat-chicken model of automation, feeding, scale, and vertical integration, but was strongest in California, the Midwest, and the eastern states, as it had been historically. In the 1950s, for the first time, flock sizes for egg-laying hens were in the thousands instead of the hundreds.

These larger egg-laying flocks became possible because of another new invention: battery cages (called so because they were placed in long rows, or "batteries"). Battery cages confined hens indoors for their entire lives in box-like wire cages. As we've seen from the *Practical Poultry Keeper*, humans had long understood and respected the egg-laying hen's need for a good nest. Until the 1950s, she had also been given space to exercise, access to fresh air and sunshine, and the opportunity to interact with her sister hens, (even if she no longer had the chance to raise any of her young). However, agribusiness researchers had been discovering that the hen's egg production is related to light's effect on the pituitary gland—more light stimulates generations of more eggs. This bit of information kicked off the drive to keep hens continually confined, where electric light could artificially lengthen their days. The move would further increase the "efficiency" of egg production. Battery cages allowed many more hens to be kept in one location with minimal attention from humans.

Unfortunately for the hapless hens, the cages were so small they allowed very little movement and, perhaps worse, they were wire-floored and devoid of nesting materials. As one can well imagine, standing nonstop on a wire mesh floor is uncomfortable in the extreme. And the cages were crowded: As many as ten hens were placed in each cage, generally giving each hen about as much space as a sheet of paper. Hens would now spend their lives confined in cramped, dusty conditions, standing on metal mesh instead of soft dirt, straw, or grass. Not surprisingly, hens' frustration and boredom caused them to aggressively peck each other. So it soon became widespread practice to cut off the pointed front portion of their beaks.

Caging hens made egg production cheaper, but, clearly, came with costs, especially for the hens. A feed dealer from this era described the "benefits" of battery cages as follows: "[W]hat they did was to organize the hens in a production line where you can use more machinery, cut way down on labor, and allow just a few people to take care of a tremendous number of birds." Professors Paige Smith and Charles Daniels, authors of *The Chicken Book*, describe the change in the following, less favorable, terms:

> Now the hen was to have a much bleaker life. Improved technology indicated that she should do as little as possible—simply eat and lay eggs. This required a controlled environment in which sun, wind, rain, bugs, worms, sprouts, and growing things had no place. For the rays of the sun a few extra vitamins in the hen's food and an electric bulb, burning ceaselessly, would be substituted; for the breeze, air conditioning; for the bugs and worms and sprouts, additives (minerals and chemicals of various kinds).

Keeping hens away from sunlight and vegetation also changed the nutritional qualities and appearance of the eggs. Yolks lost their omega-3s and took on a dull, grayish look instead of the deep yellow hue one expects from the "sun" in "sunny side up." It was quickly recognized that consumers would not accept such drab looking eggs. To correct the problem, egg producers began adding red dye to hen feed to make yolks yellow. (This is still a widespread practice today.) The nutritional shortcomings remained.

Much of the research supporting the caging of hens and other industrial practices was coming out of "land grant universities." These are institutions of higher learning that were chartered with the purpose of teaching and fostering agriculture. Each houses an agricultural experimental station and agricultural extension agents funded by the U.S. Department of Agriculture, people who are charged with supporting agriculture in their respective communities. Some of the nation's first land grant universities were Iowa State University and Michigan State University. Research from these institutions has long favored agribusiness. A national body funded by the Pew Foundation to evaluate problems in industrial animal production stated in 2008: "Industry representatives and academics agree that more public funding is needed to generate unbiased research into industrial farm animal production issues."

With the help of agribusiness research, the fate of females among the egg-laying chicken breeds become almost as unpleasant as the discarded males. Young females (pullets) are reared to nineteen weeks of age, when they are old enough to begin laying eggs. They are then placed into crowded buildings, usually in battery cages. The egg production cycle lasts about one year. At approximately one year of age, a hen is considered "spent." Somehow, USDA has decided that in the context of slaughter, birds are not animals. Therefore, it says fowl are not covered by the federal law requiring humane slaughter of all animals raised for food. "Spent hens" are frequently vacuumed up into trucks and dumped into a rotating blade chopper at a rendering plant, all done while hens are still alive and conscious.

When battery cages were first introduced, egg-laying facilities still tended to be of a relatively modest size. A Georgia outfit raised eyebrows in the early 1950s when it announced plans to build an egg operation holding 100,000 hens in battery cages. Over the next two decades, however, that size operation would become commonplace (today, egg operations with one million battery-caged hens are the norm).

Interestingly, in the middle of the twentieth century, some were already predicting that chicken and egg production's industrialization would be a harbinger of the future of all farming. A prophetic speaker at the Dixie Poultry Exposition of 1956 foresaw that "[chicken] production is the prototype of things to come in many other segments of farming."

Throughout the 1950s and 1960s, poultry production levels continued to increase although they were in excess of what was needed at home and abroad. Why? With some more digging in the annals of farm history, I learned that much of the answer lies in grain. The cost of poultry production has always been mostly in the feed. (Today, it accounts for as much as 70 percent.) Cheap and plentiful grain, therefore, tends to stimulate production.

The Truman administration had reasons, both practical and political, to support farm policy resulting in surplus grain. The most pressing reason was that at the end of World War II, Europeans desperately needed food. Millions of people were facing malnourishment and starvation. Bread was considered the greatest weapon against hunger. Thus, grain, especially wheat (and, to a lesser extent, corn), was vital in helping hungry people overseas.

After the European post-war food crisis abated, U.S. farmers kept growing more grain. Price supports encouraged increased production while, simultaneously, several technologies converged to dramatically push up crop yields. Farm machinery (including tractors), agricultural chemicals (pesticides, insecticides, fungicides, and fertilizers), and hybrid seeds were all being widely used for the first time. By the late 1950s, these technologies collectively increased the per-acre yield of wheat by 75 percent and that of corn by more than 100 percent, when compared to yields of the late 1920s.

Fertilizer was particularly important in increasing grain output. Today, corn and wheat are heavily fertilized crops. Interestingly, U.S. farmers made little use of manufactured nitrogen fertilizer for some hundred years after it first became available in 1830. At the beginning of the twentieth century, 90 percent of fertilizer used on farms was organic matter, such as manure. Then, as part of the booming wartime munitions industry of 1941, the number of anhydrous ammonia plants in the nation nearly doubled. When the war ended, affordable commercial fertilizer flooded the markets as wartime munitions plants converted to fertilizer production. The U.S. Department of Agriculture roundly encouraged farmers to make full use of agricultural chemicals to stay ahead of the curve. By 1945, artificial fertilizer use on the farm had almost doubled from just six years earlier.

The Cold War policies of the United States during the 1950s and 1960s added fuel to the grain production fire. In the global ideological struggle between the capitalist West and the communist East, U.S. leaders wanted to demonstrate to the world the superiority of their economic system. They also hoped to influence other nations with food aid. To these ends, they pushed for greater production of food staples, particularly rice, wheat, and corn. The poultry industry, which was now well organized and consolidated, quickly took advantage of the plentiful cheap grains as feed.

This U.S. grain policy also abetted concentration and consolidation of the poultry industry. When overproduction resulted in plummeting chicken prices in 1961, national agribusiness companies bought many of the remaining regional animal feed mills and processing plants at bargain prices. By the mid-1960s, the poultry industry was almost entirely under the control of big vertically integrated agribusiness enterprises. A *Wall Street Journal* article reported: "Nearly 95 percent of commercial broilers [meat-chickens] are now produced under the management of business organizations which own or control some combination of hatcheries, feed mills, processing plants, marketing services and research facilities." By the 1970s, farmers who were not under contract with agribusiness companies simply had no place left to sell either live chickens or eggs.

The "contract growing" agreements between agribusiness firms and individual farmers were invariably unbalanced due to the unequal bargaining strength of the contracting parties. That disparity enabled poultry companies to shift financial risk from themselves to individual farmers. Dr. C. Robert Taylor, Professor of Agriculture at Auburn University, has researched and written extensively about the poultry industry. He has noted that "with the growing economic power imbalance in contracting, [integrated chicken corporations] have effectively transferred income to themselves while transferring risk to producers."

Dr. Taylor's research documents how contract poultry growers find themselves on a treadmill of debt. Many contract growers take out large loans to build chicken confinement facilities, yet the buildings rapidly lose their value. "The silent and slow killer for a profitable poultry production is economic depreciation," Taylor has written.

Integrated poultry companies, on the other hand, have fared very well under the contract growing system. For example, Dr. Taylor has estimated that in 2002 the return on equity for the three largest chicken companies (Tyson, Gold Kist, and Pilgrim's Pride) averaged 16 percent for the preceding three years while the farmers raising their chickens were making no money or even losing money. Taylor also found that "profitability has decreased to the point where many contract producers have a poverty level income."

The disappearance of free markets was the final nail in the coffin for farm chickens. Ironically, although the industry kept expanding, it left no room for traditional chicken tending. As a result, chickens virtually disappeared from the barnyards of family farms. "By the mid-1950s," writes an agricultural journalist, "the old patterns of selling a few chicks to the many people on general farms, still a part of the picture at the end of World War II, was now almost gone." Whereas at the beginning of the twentieth century almost 90 percent of farms had chickens, by 1992, only 6 percent of farms had any poultry at all. The disappearance of chickens from the farmyard diminished the rural landscape while the loss of "egg money" and chicken revenue was a serious financial blow to America's struggling family farmers.

I had long known that in the second half of the twentieth century tens of thousands of farmers had lost their ability to make a living on the land. In my early childhood our eggs were delivered every Saturday from the back of a dusty pickup truck by the Schulmeister family, directly from their small farm outside of town. Some time in the late 1970s, they stopped coming. My parents explained to me that, like many other farm families, they could no longer make a living at it. When the Schulmeisters lost their farm, Mutti resorted to buying eggs at the A&P, like everyone else on our block. From that point forward, I knew exactly nothing about how or where the eggs I was consuming were produced.

I was dismayed to discover in my research that government policy had been worsening the plight of families like the Schulmeisters. A federal economic report in the 1960s was typical of the government's attitude (even if elected officials occasionally expressed concern over the loss of family farmers). The report urged an "adaptive approach" to agriculture in which farming became increasingly mechanized and it employed fewer and fewer people. For example, the report states: "Although the exodus from agriculture in the past decade . . . has been large by almost any standards, *it has not been large enough* . . . New resources (especially people) should be discouraged from entering agriculture."

For poultry industry trade groups (by this point in the pocket of vertically integrated companies) and the U.S. Department of Agriculture, it made no difference how many people were employed by farming or how well farmers lived, much less how well the fowl fared. From their perspective, there were enormous "efficiencies" gained by industrialization, concentration, and consolidation. Thanks to my poultry education, I now understood that in industrialized chicken farming "efficiency" wasn't pretty. It meant mutilated animals being raised in intensely crowded conditions, constantly fed drugs, with little human care.

Americans' consumption patterns closely track poultry farming's industrialization. In 1909, Americans ate an average of ten pounds per year of chicken and less than one pound of turkey. The amount began to rise rapidly around 1940, just as mass poultry production was really taking hold. By 2005, per capita annual consumption had risen to 60 pounds of chicken and 13 pounds of turkey—a remarkable *seven fold* increase of poultry meat consumption since the beginning of the century. The spectacular rise in consumption has been spurred partly by the popular perception that chicken is the "healthier" form of meat. In 2005, chicken overtook beef as the most consumed meat in America. Due to similar health concerns (and perhaps diminished delectability), per capita egg consumption has decreased over the past several decades and is now actually lower than it was at the outset of the twentieth century.

THE GOOD SHEPHERD OF BIRDS

Thankfully, a few independent traditional poultry farmers remain. Several years after starting my inquiry into the poultry industry, I would meet one such farmer, a memorable teacher about all things fowl. Frank Reese raises turkeys, chickens, and ducks in Lindsborg, Kansas at the Good Shepherd Turkey Ranch. I first met Frank at a conference in Washington, D.C., when he and I spoke on the same panel. Later, I had the pleasure of visiting his wonderful farm. Frank immediately impressed me with his passion and depth of knowledge. "Genetic diversity is my big thing," he told me in a conversation after the conference. Modern poultry flocks have a variety of health problems and are more susceptible to diseases because of, in addition to the conditions in which they live, the way they've been bred by industrial operations, Frank explained. He ardently supports the alternative: hardy traditional poultry, often referred to these days as "heritage breeds."

Frank's poultry passion has deep roots. His great grandparents came to Kansas shortly after the Civil War, and the family has been farming in the region ever since. Frank grew up raising turkeys and always enjoyed being around them. His turkeys descend from bloodlines developed by renowned poultry breeders of the early twentieth century. "I just never switched to industrial animals," he explained to me. Older chicken and turkey breeds were spangled, barred, white, black, red, buff, blue, and "Columbian," a feathering pattern with white for most plumage and black in the neck, wings, and tail. Such colorful and interesting birds have always been found on Frank's farm.

"About ten or fifteen years ago, I saw that we were in real trouble," Frank told me as we walked the yards of his farm together. "All the folks I knew who understood how to maintain the older breeds, they were getting old and dying off." At that point, Frank says, he began to feel a pressing responsibility to preserve traditional poultry breeds.

The problem is that turkeys, chickens, and laying hens raised by industrial operations are each of a single (always white) breed (such as Broad Breasted White turkeys and White Leghorn laying hens). Industrial production aims to bring about uniformity, so, by design, the birds are genetically unvaried. At the same time, breeding for certain characteristics (like high egg production) has been pushed far beyond advisable limits. With turkeys, the obsession has been big breasts. The anatomy has been made so extreme that turkeys now have difficulty with basic bodily functions. "The industrial birds have been bred in such a way that they can't really survive in nature—they can't even mate," Frank says. "Everything has to be done with artificial insemination. It's just crazy."

While the 1943 manual *Poultry Breeding Applied* noted that

artificial insemination for chickens had first been developed in 1935, artificial breeding of turkeys was still rare. The inability to reproduce is a phenomenon of only recent decades. Until around the middle of the twentieth century, fowl remained perfectly capable of mating on their own. Today, artificial insemination is universally practiced at turkey operations.

Beyond their inability to reproduce, industrial breeding has made poultry bodies generally unfit for survival. "Both turkeys and chickens have really lost their ability to live," Frank told me. "They can't even walk very well—their bodies are just not sound."

Then, with evident pride, he quickly added: "My birds cannot only walk—they can *run*. And they can *fly*, too!" Frank's heritage birds spend lots of time on pasture, breed naturally, and eat only natural, drug-free feeds—including plenty of bugs they find from their own hunting and foraging.

Frank's breeding toms (males) and hens live in groups according to breed, including the Bronze, the Narragansett, the White Holland, and the Bourbon Red. Each flock has a wooden barn with perches, nesting boxes, and plenty of space. Adjacent to each barn is a large, vegetated yard with trees, brush, and small shelters. As I wandered through his barns and yards observing his birds pecking the ground and interacting with one another—males displaying the full glory of their fantastic feathers for the females, I realized for the first time that turkeys are beautiful. I could finally understand why Benjamin Franklin famously wrote that in comparison to the bald eagle, the turkey is "a much more respectable bird," admiringly calling it "a true original Native of America" and a "bird of courage."

Frank and his farm have gained impressive national attention, largely because of the extraordinary quality of the food he produces and how greatly it differs from industrially produced poultry meat. In 2001, Marian Burros of the *New York Times* was looking for "the best tasting turkey in America" for Thanksgiving. After searching the country, she settled on Frank's. Later, he ended up on Martha

Stewart's show and was even chosen as the Person of the Week by ABC News. Frank has many devoted fans in the food world. Hopefully, he will one day have as many farmer disciples.

POULTRY INDUSTRY LESSONS

During my first months on the industrial animal operation campaign at Waterkeeper, I received a bunch of phone calls from activists troubled by chicken or egg operations. Each of them described industrial poultry complexes in their communities that were causing awful stenches and polluting water with nutrients and bacteria. A Sierra Club lawyer called to discuss a case he wanted to file under the Clean Air Act. The calls were hardly surprising when one considers that a typical building for meat chickens produces more than 200 tons of poultry litter annually. People were hearing about Bobby Kennedy's interest in the issue and were soliciting our organization's involvement and advice. We planned to focus first on the pork industry. Yet I devoted time to these calls because we intended to eventually press for reform in every part of the food animal industry that was causing pollution.

My investigation of poultry farming's transformation had been a revelation in many ways. For one thing, it was now clear to me that the disappearance of the family farm was not the result of some unstoppable natural force like wind wearing down mountains. Industrialization, rather than being inevitable as is often believed, had been aided by public policy and stimulated by individual entrepreneurs looking to maximize profits. Much of this "progress" had occurred at the expense of farmers and animals.

Industrialization had made its boldest advances at those moments in history when family farmers were suffering setbacks—when they were especially vulnerable. As Southern farmers were watching their cotton lose value in the 1930s, they were talked into raising chickens on contract. As Delmarva farmers faltered en masse in 1946 after the U.S. government cancelled its wartime contracts, farmers reluctantly converted to contract production to avoid losing their farms. And as chicken prices precipitously fell in 1961 due to overproduction, large companies further consolidated the industry by swallowing up regional poultry processing operations.

I would soon learn that this pattern of industrialization would repeat itself in hog farming. When farmers in tobacco growing regions found themselves on the ropes, pork agribusiness companies offered them attractive sounding deals to raise pigs on contract. With this new understanding of farming's industrialization, I turned my attention to pork.

CHAPTER FOUR

Facing off with Big Ag

A NATIONAL CAMPAIGN
FOR REFORM IS BORN

On a sunny mid-November afternoon, I carried six packets for certified mail to the post office up the road. It was my first formal action against the meat industry and I was elated. Each envelope contained a letter officially notifying a corporate hog operation of our intentions to sue. The court pleadings were not yet complete, but I'd gathered enough facts and law to know what we would be alleging. The certified letters—a mandatory first step in filing suit under federal environmental laws—accused the facility operators of contaminating air, water, and soils with hog manure. Our campaign was finally in motion.

By this point I shared Bobby's mounting sense of urgency to stop, or at least slow, a plague spreading across America. I was now fielding daily calls and emails from people in Oregon, Minnesota,

Alabama—any place with confinement agriculture. They had heard about Bobby's interest in industrial animal facilities and wanted us to know their individual stories of pollution, injustice, and animal cruelty. It had begun to feel like I was sitting in command central as a natural disaster was sweeping the country. Yet this scourge was entirely man-made and deliberate, making it all the more vexing.

My initial caution about filing litigation melted under the weight of the evidence we gathered, and for weeks I'd worked feverishly to finish our pleadings. It was taking me much longer than I'd expected. I wanted the facts and law we cited to be unimpeachable. To prepare our cases for court, Rick continued collecting photographic evidence and field data while I searched law books for relevant cases, statutes, and regulations. Because, true to my nature, I wanted to be absolutely thorough, and because we were charting new territory, it was slow going. Bobby joked that it was the longest he'd ever taken to get out a notice letter.

Our cases would be among the first to apply America's major environmental laws to agriculture. Hog facilities were breaking the Clean Water Act, our notice letters alleged, because while they had no pollution discharge permits, they were directly and indirectly contaminating groundwater, streams, and rivers. And they were violating the Clean Air Act and federal solid waste law (known as RCRA), we argued, because they were polluting air and soils without air or solid waste permits.

These would be "citizen suit" enforcement cases. Knowing that polluting industries would invariably pressure politicians and government agencies to refrain from applying environmental laws, Congress wisely empowered ordinary people to enforce environmental laws by acting as plaintiffs in such enforcement actions. Our plaintiffs were the Waterkeeper organization and clean water advocates who lived in the area affected by the pollution. The defendants were all from the list of hog facilities Rick and I had assembled in North Carolina.

The enforcement lawsuits were the first element of a multifaceted campaign against industrial animal production, the shape of which was emerging from our regular internal strategy sessions. It had become clear to us that all levels of government were, at best, anemically enforcing environmental laws against agricultural operations. No state even required the pollution permits mandated by federal statutes. As Bobby put it, "They're getting away with this stuff because of the meat industry's influence in Washington." We believed that bringing enforcement cases could have the dual effect of addressing specific pollution problems and prompting federal and state environmental agencies to heighten their attention on animal facilities. That made enforcement actions key to our plan.

As a second part of our strategy, we had committed ourselves to assisting activists. In communicating with citizens around the country, a lot of whom were farmers, it had become obvious that many were unconnected to one another and had limited access to empirical public health and environmental data. Information is power. We wanted activists to have the power to bring about reform. So helping activists organize and get good information was another priority.

Finally, we wanted our campaign to help establish that industrial methods for raising animals were both undesirable and unnecessary. I felt especially strongly about the need for this aspect of our overall strategy. "Let's not be just another group attacking the meat industry," I urged in our discussions, "we should find ways to promote good, viable alternatives." Rick and Bobby completely agreed. But what was the best way to go about it? We all knew that politicians and regulators were captives of the powerful agribusiness and food industrial complex, making reform by legislation or regulation highly unlikely.

Influencing consumer demand, therefore, seemed a much more promising means for changing the status quo. But consumers weren't agitating for an overhaul of their food system. Few had a clue about the history of their food—least of all meat, eggs, and dairy—before it reached their plates. This meant we needed to figure out ways to speak directly to the food buying public about what was happening at industrial animal operations. Media coverage of our lawsuits could be our first communication to that audience. I felt we also needed to begin looking for good alternatives to industrial foods to recommend to consumers.

Alongside these various efforts, Bobby was focused on a still more aggressive legal strategy. It involved using the law to go after the corporate meat industry on a national level. His friends include dozens of successful plaintiff's lawyers, and Bobby intended to enlist their help in this effort.

One of them was Jan Schlichtmann, an attorney portrayed by John Travolta in the movie *A Civil Action*. Years earlier, Jan had been financially ruined by tenaciously pursuing an unsuccessful toxic tort class action against a major chemical company. Many lawyers would surely shy away permanently from such work after the painful experience. But he got back on his feet, dusted himself off, and joined a prestigious plaintiff's firm, hoping to pursue more environmental cases. Jan's personal perils actually seem to have strengthened his resolve to go after polluting industries, especially those hurting the disenfranchised poor and powerless.

Accompanied by Charlie Speer (the lawyer I'd met in Missouri) Jan came to brainstorm at our office in New York. A tall, lanky man with dark eyes, a thick shock of snow-white hair and a fluffy moustache, he bears absolutely no resemblance to John Travolta, and I immediately liked him much better than the film's character. From my interactions with him, I suspected Jan was motivated by both sincere moral outrage and the prospect of hitting the next big jackpot. At the end of a few hours of discussion around our conference table, Jan, Charlie, Bobby, and I were satisfied we should collaborate. "This industry is really the perfect target," Jan said with a wide grin as we closed the meeting.

That meeting was the first in a series with high profile plaintiff's lawyers. Every one of the lawyers Bobby approached enthusiastically signed on. The group was eventually made up of about a dozen accomplished attorneys and dubbed the "superstar lawyers" by Bobby. It included, among others, Jan, Charlie, Richard Middleton (a past president of the National Trial Lawyers' Association), and Steve Bozeman (one of the plaintiffs lawyers suing the tobacco industry depicted in the movie *The Insider*). They had the skills and resources to challenge big agribusiness like no other group before them. Within weeks, each anted up $50,000 for a war chest to fund their litigation. We were giddy with the potential.

But the slow pace of the group's deliberations soon splashed some cold water on our initial enthusiasm. Among the lawyers, opinions differed about what legal theories to pursue and where to file the case. Originally, Bobby had hoped to simultaneously launch a battery of federal lawsuits—my six federal enforcement actions along with the cases that would be prepared by the outside lawyers. The way things were going, however, it became obvious that would not happen in the near future. "We'll go ahead as soon as you're ready," he alerted me privately, "just get your cases done." Bobby was anxious to announce our reform campaign to the world. Now I began to worry: Would the world pay attention?

Once the enforcement action notice letters I had prepared were in the mail, Rick and I started planning press conferences to be held on the same day in Raleigh and Washington, D.C. Between us, we were calling and emailing literally hundreds of reporters, on both national and regional beats. Bobby and his group of superstar lawyers would be declaring war on the corporate animal industry. Lots of reporters were intrigued, although they stayed noncommittal about attending our press events.

We wanted the press conferences to unveil a new coalition of unlikely allies. Our plan was to gather animal activists, farmers, and environmentalists. I'd also been working with faith-based organizations that were advocating for traditional agriculture as a way of producing food that was kinder to all aspects of God's creation. We believed that only if these diverse, smaller groups joined forces could we effectively counter the considerable political power and resources of agribusiness. Individually, we were each David standing up to Goliath. United, we would have the strength to mount a formidable campaign for reform. Coalition representatives would stand beside Bobby at the press events as he announced that industrial animal facilities could no longer operate with impunity.

The idea was good in theory, but it presented the practical difficulty that these different camps had long regarded each other with profound suspicion. Animal advocates (often, in my experience, lumped together by farmers as "those PETA people") and environmental activists make many farmers terribly nervous, and vice versa. No farmer likes outsiders telling him how to run his farm, least of all people who don't think he should be raising animals at all. Many in the animal protection community view farmers as perpetrators of suffering, or worse. Increasingly these days, animal groups take the position that all animal-based food or fiber production is immoral and advocate for a human existence entirely devoid of meat, dairy, eggs, fish, wool, leather, silk, or even honey. Some environmentalists consider farmers and ranchers inherent enemies of the ecosystem.

Aware of this complicated history, we aimed to overcome past tensions by focusing on the common ground of stopping the rise of industrial animal operations. Bobby's stature made it possible to pull together these strange bedfellows. I believe it was due to his involvement that we were able to swiftly galvanize major animal, environmental, and religious groups.

The farmers, however, as tends to be the farmer way, were showing more reticence. Bobby decided to phone the president of the National Farmers' Union, Leland Swenson, and personally appeal for his support. I sat in on the call. "These big meat companies involved in the livestock business aren't real farmers, and they don't care about farmers," Bobby asserted with passion. "Real farmers aren't even being given a fair chance to compete anymore. Given a level playing field, every good family farmer in America could out-compete these big animal factories." Everything Bobby was saying seemed to be music to Swenson's ears. At the close of the call he agreed, although perhaps still with some reservations, to stand beside Bobby at the Washington, D.C., event.

On the eve of our press conferences I flew to Raleigh where Rick Dove again met me at the airport. As he and I dealt with what seemed like hundreds of last minute details, we chatted on our cell phones with a steady stream of reporters. We were beginning to feel cautiously optimistic we would have a good turnout.

As dawn broke the next day, December 6, 2000, I was racing to the airport in a rented car to collect Bobby and Jan Schlichtmann, who were scheduled to arrive on the first flight from New York. Luckily, the flight landed on time and they soon appeared at the curb. Bobby and Jan hopped into the car and I sped toward downtown Raleigh. We got there about five minutes before the press conference was scheduled to begin. It was thrilling to find the room jammed with farmers, environmental activists, and reporters, including three television crews. A Kennedy declaring war on one of the state's largest industries was news indeed.

Bobby confidently strode straight to the front of the room and grabbed the sides of the podium. As always, he spoke without notes. He began by taking direct aim at industrial animal operators, accusing them of polluting, treating animals inhumanely, and driving true family farmers out of business. Then he called to task public officials for allowing this to happen on their watch. "Federal environmental prosecution against the meat industry has effectively ceased," he asserted. "Congress has eviscerated the Environmental Protection Agency's enforcement budget while the political clout of powerful pork producers has trumped state enforcement efforts," he pronounced to the crowd, punctuating his remarks with an accusatory pointing finger. "This collapse of environmental enforcement has allowed corporate hog factories to proliferate with huge pollution based profits."

Jan Schlichtmann stepped up to the podium next. "I am looking forward to facing off with an industry that has been injuring people for years," he declared. "Our objective is to civilize this industry." When Jan had finished speaking, reporters starting throwing questions at him and Bobby, most of them tending toward the potential economic impact of our lawsuits. "Look," Bobby responded, "responsible waste management is part of every business' costs. When you're handling your waste in a way that breaks that law, you're competing unfairly. All this will do is level the playing field."

Before we hurried out the door to catch our plane to D.C., Bobby stopped to shake the hands of several of the farmers and activists. Farmers dusty from morning chores had driven in pickups from all over the state to thank him for standing up to an industry that had been bullying them for years. They were black, white, men, and women—all "beef jerky tough," as Bobby said later. Several came from families who've occupied the same piece of earth for generations. I was astonished to see tears flowing down some of their weathered faces. Bobby was visibly moved. "Thank you for taking the time," one of those farmers said simply. He seemed deeply touched that someone of Bobby's prominence wanted to be their standard bearer.

Three hours later we landed at Washington National Airport, giving me just enough time to get the room at the National Press Building set up and make sure everyone was present and prepared. As I checked to see that our full cast was in attendance—all our coalition organizations and several of the plaintiff's attorneys—I kept an anxious eye on the door to track who was showing up from the press. The room slowly filled to capacity with reporters.

Bobby and Jan again launched the session with a string of impassioned accusations against animal agribusiness. Following them, National Farmers Union president Swenson took the podium and announced his organization's support for our campaign. "Industrial hog production is not farming," he stressed. My sentiments exactly, but so much more meaningful coming from him.

Then a petite, fair-haired woman stepped forward. She pulled down the microphone and cleared her throat. It was Diane Halverson of the Animal Welfare Institute, a respected animal protection organization. "The treatment of pigs in animal factories is nothing short of barbaric," she began softly but with determination. Diane presented a distinct contrast to the alpha males preceding her and the room became perfectly still as she continued. "The time has come to revitalize the culture of animal husbandry that respects the animals." With each sentence, her delivery was a bit bolder. A few minutes later, speaking in a full, clear voice, Diane closed with a quote from Nobel Prize–winning humanitarian Albert Schweitzer: "We need a boundless ethics which will include the animals also."

In about two hours, it was all over and felt like a success. The room had been full of clicking cameras and eager reporters brimming with questions. Now we would wait and watch for what media coverage actually came of it.

We packed up our easels and extra press packets and rushed out the door to catch a plane back to New York. On the flight I ordered a cold beer, which went down like nectar. Leaning back into my seat, I closed my eyes and felt a satisfied smile crossing my face. Suddenly I remembered with a start Bobby's announcement at the press conferences calling for "a national summit on hog farming." I realized with a wave of fatigue I'd have to tackle that task without delay. I had my marching orders and there would be no resting up from this event.

RALLYING THE TROOPS

With the dust from our press conferences not yet settled, I threw myself into planning the summit announced by Bobby. As my long work day stretched still further, I was becoming accustomed to having breakfast, lunch, *and* dinner at my desk, rarely making it back to my apartment before 10 p.m. It is no overstatement to say my job had become my life. The rigorous exercise routine I had followed for years was now nothing but a dim memory. And the notion that New York would proffer a bevy of interesting, eligible young men was shattered by the reality that I was never in a position to meet any of them, let alone go out on a date. A strong sense of purpose riveted me to my desk. Had I not been loving my work, I'd have been miserable.

The summit I was now planning would be a full day affair, gathering experts and activists from around the country in New Bern, North Carolina, near the banks of the Neuse River. Our goal for the event was ambitious: People would come together, share information then depart as a vigorous national network collaborating to reverse animal agriculture's industrialization. Feeling that the event needed to be as inspiring as informative—much more like attending a rousing church service than a symposium—I began referring to it as our "tent revival meeting." The program would be sprinkled with potent speakers then capped off with a barnburner by Bobby.

Meanwhile, the meat industry was strongly and swiftly striking back at us in the weeks following our press conferences. Ironically, its response was bringing additional publicity to our campaign. Rather than denying our charges against industrial meat production, the industry publicly criticized us personally, mainly along two themes.

Industry attack line one: "These are a bunch of vegetarian activists whose true motivation is to permanently wipe out animal farming and meat eating." A leading agribusiness figure announced that we were "the twentieth century's greatest threat to agriculture." This was all tremendously flattering, but laughable. Bobby's a hunter and both he and Rick are avid fishermen. They and everyone else in our organization (except me) were committed carnivores with absolutely no interest in giving up their bacon.

Industry attack line two: "These are just naïve, ill informed East Coast lawyers who know nothing about agriculture." Now, that one hit closer to home. None of us had farming backgrounds. Admittedly, we had much to learn about this complex subject, something I was realizing several times a day.

Both of these industry attacks on us could best be answered, I felt, if our actions—not just our words—supported farmers raising animals the right way. Partly for that reason, we chose to name our meeting the "Summit for Sustainable Hog Farming" (as opposed to, say, "Summit Against Factory Farms"). I began asking around among animal welfare activists and environmentalists about good farmers we could promote as models at the Summit. Everyone I queried agreed that there was very little "gold standard" livestock farming happening today and even fewer companies committed to buying exclusively from such farmers.

Just one name kept surfacing: Niman Ranch. It was a California-based natural meat company supplied entirely by a nationwide network of traditional farmers and ranchers following strict animal husbandry standards. None of the Niman Ranch pig farmers used confinement buildings, liquid manure systems, sow crates, or any other practice we condemned. And it was the only organization in the country whose pig farmers all followed animal handling protocols crafted by the Animal Welfare Institute. Rick and I quickly decided that we wanted Niman Ranch centrally involved in the Summit.

Through an intermediary acquainted with him, I invited the company's CEO and founder, Bill Niman, to speak. He immediately accepted and volunteered a donation of several hundred pounds of pork for the event's dinner. But a twist of fate ended up forcing him to cancel a couple of weeks later. I would not meet or speak with Bill Niman for another six months. In his stead, he sent a senior company officer along with one of its pig farmers. He also delivered on the pork.

I was delighted we'd be serving Niman Ranch pork chops at the event. The Summit would be the world's first gathering of "vegetarian activists" with livestock farmers as featured speakers and pork as the main course. I could think of no better way to answer the industry's specious claim that this was a meeting of wild-eyed vegan radicals plotting the demise of meat eating.

About half of our time getting the Summit ready was spent recruiting people to attend, involving ample begging and cajoling. Rick called every activist he knew (some several times), pressuring them to commit to coming and bringing at least one other person. To encourage attendance we also set the entry fee for the whole day's event, including breakfast, lunch, and dinner, at just $25 (far lower than our actual cost).

In spite of these efforts, Rick and I both remained preoccupied with concerns about attendance. Filling the hall was crucial to spreading our gospel and to creating an energized atmosphere. About three hundred people signed up in advance, which would be a respectable turnout. But we had much grander plans—we wanted to fill the convention center. That would take closer to a thousand.

I vividly remember the morning of the Summit, January 11, 2001. It was cold and partly cloudy; roads were clear and dry. Around 7:30 a.m. I was nervously looking out the convention center's glass front doors as the first busload of North Carolina Summit-goers arrived. Minutes later, cars, a van, then another bus pulled into the parking lot. Yet another bus arrived, this one from Virginia. People were now streaming into the hall. Among them were Terry Spence and Scott Dye from Missouri, and Diane Halverson from Minnesota, who had all become our close allies and friends. By 8:00 a.m., the convention center was bubbling with activity.

A large, white banner across the main entrance announced to the citizens of New Bern what was happening inside. Dozens of reporters had come to tell the rest of the world. Among them were national correspondents from NPR, the *New York Times*, and numerous agricultural journals, and about two dozen local and regional reporters, including several television correspondents and cameramen.

Shortly after 8:00 a.m. Rick and I took the stage to welcome the crowd. People were greeting old friends and forming new acquaintances. The air seemed to hum with excitement and anticipation. As I stood before this energized throng, my heart swelled. It looked as though the Summit would be everything we'd been hoping for. The crowd settled down as Rick and I began to speak. "Welcome to the first Summit for Sustainable Hog Farming," I opened. "I cannot tell you how happy it makes me to see you all here in this room. Today we are taking our first step together toward restoring a sensible way of raising hogs, farming that's good for farmers, animals, the land—and everyone eating pork." The audience broke into approving applause. Rick then spoke a few words of welcome.

After we stepped down, a parade of citizens and scientists began taking the stage to share personal experiences or describe their research. One of the day's most memorable speakers was Karen Priest, a mother and homemaker. Like most activists I've met, she first got involved when she felt she needed to defend her family and her community. Karen is a dignified, pretty woman with short, highlighted hair and a voice that's honeyed with a soothing southern accent.

"My story starts about 15 years back," she told the attentive audience. "Me and my husband had our eye on this one house for years. I'd just fallen in love with it. The house is built from this local sandstone and sits on a gorgeous 12-acre lot. It's set back from the road, surrounded by big ol' trees. We started talkin' about it as our 'dream home.' Then, one day, it went up for sale. We jumped at the chance to own it and made an offer that same day. The house needed a lot of repairs, but we were okay with that. After we bought it, we spent two years working on it, evenings and weekends, while staying at our old place. When we could finally move in, we thought we'd found paradise," Karen recounted. Those happy days would turn out to be all too few.

"A couple months later, our neighbor to one side brought in a bulldozer and started moving dirt around. It was the first notice we had that he was puttin' in a hog factory and manure lagoons. Here in North Carolina, there's no requirement to tell that to the neighbors. It would become a Murphy Farms operation with 7,300 hogs. Naturally, we didn't like the idea, but we figured there was nothin' we could do because it was their property. Then, two years later, I found out that on the other side, a Prestage Farms operation would be goin' in. It was goin' to have 5,800 hogs. We'd be surrounded."

Because Karen and her neighbors now knew all too well what it was like to live near a hog confinement facility, they got themselves organized. They began holding meetings with each other and attending sessions of the county commission. But the local board showed little interest in the neighbors' concerns. As Karen explained, "We have hog growers on our county commission." Also to no avail, Karen and her neighbors tried to get the newspaper to cover their concerns. "The Bladen Journal," she explained, "seems favorable to the hog industry, and no public official dares speak out against them." The Prestage Farms operation was soon up and running.

Karen and her group of neighbors were not easily deterred from their cause. They began contacting the state's Department of Water Quality whenever odors invaded their homes, which was frequently. "It's sad to say, but I felt the love of our house jes' slippin' away," Karen continued. "We have to sleep with our windows closed on summer nights. I can't hang my laundry outside. I'm always nervous about havin' people over because I worry about the smell startin' up. It's been especially bad for my kids. They're afraid to have their friends over. They're embarrassed that their house stinks. It gets me so frustrated and angry that we have to live like that in a house I'd once thought of as our dream home."

Karen then told the audience how one day she decided to throw caution to the wind and host a party for her young daughter's birthday. As she began to speak about it, a slight tremor crept into her voice. "The kids were playing in the yard, and everything seemed to be goin' fine. But then I noticed the hog smell rollin' in. Kids were comin' up to me and complainin'. One of my daughter's friends asked to use the phone. She went in and called her parents.

She told them it stank real bad at our place and that she jes' wanted to go home—could they come pick her up early. It was humiliatin' for me and my daughter, both. That was the last time we ever had a birthday party at our house."

Karen and her neighbors spent years urging the state to take some action about the smell. Finally, the state found one of the facilities to be in violation of the state's odor rule and directed the operation's owner to abate it. She was frustrated that he had not yet complied. "He was ordered to plant bufferin' trees. He said he has 'not been able to secure the trees.'" Karen finished speaking and sat down. I could see that many in the audience had been touched by her story.

Later, another particularly remarkable speaker took the stage. Don Webb is a tall, broad shouldered man with the stage presence of Billy Graham and a booming, intensely southern flavored voice to match. When Don spoke—I should say *preached*—at our Summit, he was already well-known in the state for his years of activism against confinement hog facilities. Even more compelling than Don's delivery was his story. And no one tells a tale like Don Webb.

Don recounted how one day he decided to supplement his teaching salary by raising some hogs. By this point, confinement hog raising was dominant in North Carolina and was being heavily touted as "*the* future of hog farming" by land grant universities and agribusiness. Don decided to follow their advice and use the most "modern" methods. He dug a huge hole for liquefied manure, erected a metal hog shed with a slatted floor, and filled it with pigs. For a while, things seemed to be going alright, and Don felt he'd made a wise investment. Then the place started to smell something awful.

"I kep' callin' the company who'd sold me the equipmen' an' askin' them how I could make it stop stinkin' so bad," Don recalled for the audience. "The lady I talked to on the phone tol' me, 'Ya'll go out into that lagoon an' add this chemical. It'll fix everythin.' So, I put my boat in that cesspool an' rowed it way out to the middle of that slop. Then I poured that chemical in there and stirred it up like a giant pot o' chili with one of them oars. Tears were comin' outta my eyes the whole time from the fumes. But I'll tell ya'll this—it did not make one dang bit o' difference. Truth be tol'—it made it *worse*! Soon I was smellin' up the *whole town*!

"Anyway, I wasn't sure what to do about it," Don continued, on a roll now. "I had put a buncha money into the buildin' an' the lagoon, so I couldn't jes' up an' quit raisin' hogs. Then one day I was at the grocery store. A nice young man I'd known a long time come up to me. He's a very polite person and kinda shy but I could see he had somethin' to say. I asked him what it was. 'Well, Mister Webb,' he said, 'I don't mean to be rude, but it's 'bout your hog lagoon. My family and I, we were jes' wonderin' if, if there's any way you could stop it from stinkin' so bad. It smells up our whole house.'"

That was Don Webb's Come-to-Jesus moment. Shortly there-after, he closed down his facility and gave up confinement swine raising for good. And when a large industrial facility was proposed near his land, he became a devout contra-industry activist.

To say Don has been disillusioned with industrial hog raising would be putting it mildly. He despises it. He's seen it's ugliness from inside and out. And he does not mince words. "Those boys over at the pork council they'll tell you this is the only way to raise hogs nowadays. But that's a load a crap! It's jes' the way they can make the most money at it." Don's face was now crimson and his voice had grown louder and animated with anger. Indignantly, he cried: "What they're doin' to the *animals*, and what they're doin' to their *neighbors*—*it's a sin*!" He slammed his fist on the podium as the crowd roared in lavish appreciation. And I was loving every minute of it. Now it was really starting to feel like the evangelists' meeting I'd been picturing all along. I pitied the poor soul who had to follow Don on the agenda.

Our roster of speakers also included several university scientists whose presentations were admittedly a bit drier than Don's but were full of crucial information. In planning the Summit, Rick and I had been well aware that university researchers would be pressured to stay away by the pork industry. Nonetheless, we contacted about ten university-based scientists who were doing significant research into the public health or environmental effects of industrial animal production. We considered their findings vital to the Summit program.

Dr. Joann Burkholder, the *Pfiesteria* discoverer, had immediately agreed to attend when contacted by Rick. Her quick, definite acceptance of our invitation affirmed once again that she was a person who would not be intimidated by the lawyers and spin doctors of agribusiness. Using slides to illustrate, her talk at the Summit described how spiking nutrient levels in North Carolina's waters coincided with the rise in industrial animal production. And she addressed some of the many environmental and public health risks that result from such pollution.

Dr. Steve Wing, a sociologist from University of North Carolina, had also unflinchingly agreed to participate in the Summit. He presented the audience his research, which documented that people living near hog factories suffer higher than normal levels of depression, anxiety, nausea, and sore throats. His research also found that industrial hog facilities were disproportionately located in communities with high levels of poverty and high proportions of minority residents. Dr. Wing then described what happened when he completed his study. "Lawyers from the hog industry paid me a visit," he recounted. They were threatening to sue. Undeterred, Dr. Wing bravely proceeded to publish and speak about his findings.

Another speaker with a similar story was Dr. Melva Okun, a public health expert. She had written her Ph.D. thesis at University of North Carolina about human health problems related to hog confinement facilities. It detailed waterborne and airborne diseases from hog operations as well as injuries, such as chronic lung diseases, from exposures to their noxious gases. She was also employed by UNC. After turning in her dissertation, people in the hog industry contacted her university employer and raised a stink. She feels that she lost her job as a result.

By the end of the day, almost a thousand people filled the hall. When all the citizens and scientists had finished speaking, Bobby ascended the stage. The crowd eagerly awaited his words. "When Rick Dove kept talking to me about hog factories, it took me some time to catch on," he began. "But after a while I realized that just as GE had polluted the Hudson River then pulled up stakes and abandoned the communities it had contaminated, Smithfield was doing the same here in North Carolina." When Bobby concluded, the audience jumped to its feet, thanking him with a long, rowdy standing ovation.

The invitations we'd extended to speak that day were not all accepted. Dr. Viney Aneja, who had documented massive nitrogen air pollution from animal confinement facilities, did not make it to the Summit. We suspected it was from industry pressure. Rick had called to personally invite Dr. Aneja and initially he had enthusiastically signed on. But a few weeks later, with effusive apologies, he called Rick to back out. By the time he withdrew, our meeting was a hot topic in the pork industry. Trade association newsletters were encouraging their members to picket the event.

Likewise, North Carolina's director of public health, Dr. James McBride, seemed to be feeling heat from the pork industry. He had been actively warning citizens about industrial animal operations releasing dangerous pathogens to the air and water. Among other actions, Dr. McBride was posting "No Swimming" signs at beaches around the state due to high levels of bacteria and outbreaks of the toxic *Pfiesteria* microbe. But the industry was lashing back, adamantly protesting its innocence and claiming it was suffering unfairly under this overzealous bureaucrat. Dr. McBride told Rick, "I would love to speak at your meeting, but I just can't under the circumstances."

Since those days organizing our first Summit, I've met more than twenty scientists who have described having their work chilled by animal agribusiness. They function under constant overt or indirect pressure to refrain from researching, writing, or speaking on the industry's environmental or human health impacts. The situation is the inevitable outgrowth of a system in which agribusiness provides the bulk of funding for university-based agricultural research. Not surprisingly, agribusiness has no interest in funding studies that might reflect poorly on itself. As to those rare instances where such studies do get completed, the authors then must often struggle to write or talk about their findings. Inevitably, this dynamic virtually dictates what research does and does not get done, and published, on the topic of farming in the United States. Study of alternatives to industrial systems for animal raising is almost nonexistent.

Unfortunately, censorship of researchers, even by our own government, is neither new nor rare. While seeking presenters for our North Carolina Summit I learned of an incident where the government was already censoring important agricultural research many years earlier. I encountered a man, by then in his nineties, who gave me his own firsthand account. In the early 1940s, the U.S. Department of Agriculture decided to study the effects of agriculture's industrialization, then an emerging phenomenon. It hired a bright, young sociologist named Dr. Walter Goldschmidt to look into the question.

After months of thorough research, Dr. Goldschmidt submitted a report that reached unequivocal conclusions on the impacts of industrialization. He found that while traditional farming greatly benefits communities, industrial agriculture is detrimental. Traditional farming contributes to vigorous citizen participation, good schools, low crime, physically healthy populations, and a high quality of life. Industrial agriculture, however, has the opposite effect in every category. Hopeful that his research would help shape federal policies to favor traditional farming over industrialization, Goldschmidt turned in his report.

But the emerging agribusiness industry learned of the study's findings and complained bitterly to USDA. In response, Dr. Goldschmidt's bosses swiftly shelved the study, shunned him, and continued promoting the industrialization of food production. One wonders why it commissioned the study at all. Dr. Goldschmidt went on to become an esteemed sociology professor at UCLA, but decades later he still vividly recalled the frustrating experience. "I hope you'll tell the story of what happened to that study," he said to me over the phone.

A year after our North Carolina Summit, I personally witnessed another incident of USDA censoring one of its own researchers when the findings cast an unfavorable light on agribusiness. I was organizing our second Hog Summit, this time in the epicenter of hog farming, Iowa. One scientist we particularly wanted on the program was Dr. James Zahn, a microbiologist with USDA's Ames, Iowa national swine research facility. Through the grapevine we knew he had conducted cutting edge research establishing that industrial hog facilities can release millions of antibiotic resistant bacteria into the air, which can float miles from their sources. The study raised the specter of a broad public health threat from animal confinement facilities.

Needless to say, the pork industry was mightily displeased by Dr. Zahn's conclusions. Under industry pressure, USDA discouraged him from speaking openly about his studies although they had all been publicly funded. According to Dr. Zahn, on more than ten occasions he was barred from presenting his findings. His superiors then forbade him to speak at our Iowa Hog Summit after they had been contacted about it by the Pork Council, the industry's largest trade group. The back and forth between pork industry representatives and Zahn's bosses was all documented by a chain of correspondence. That seemed to be the last straw for Dr. Zahn. He resigned a few months later.

A BIG VICTORY

The successes of our North Carolina Hog Summit and our press conferences infused our campaign with an intense forward momentum. Both events were widely covered by the media. During this period, our press clipping service collected more than a thousand articles about our campaign from newspapers and magazines around the country. We were becoming known to organizations and activists everywhere as a central clearinghouse on industrial animal production. Much of my day was now spent responding to inquiries from citizens and lawyers. I took in their stories and willingly shared everything I knew about how to organize against and sue industrial animal operations.

A month after the North Carolina Hog Summit, I finished and filed our first two federal legal complaints, both against North Carolina hog facilities owned by Smithfield Foods. By then I was deeply embroiled in all facets of our campaign, so the event struck me as less noteworthy than the moment I'd mailed our notice letters a few months earlier. But it was certainly another important step forward in our efforts.

Smithfield, of course, had a crowd of high-priced corporate lawyers defending itself. They responded by filing a motion and thirty-three-page brief arguing for dismissal of all our claims in both cases. I had seven weeks to respond. As plaintiffs, this was a pivotal moment for us: Losing the motion would be fatal to the cases and a major psychological setback to the entire movement for sustainable animal agriculture. While pushing ahead on all fronts of our strategy, I made researching and drafting our answer to the dismissal motion my highest priority.

On the day it was due, I filed a thirty-two-page response. The opening paragraph of the Factual Background section read:

The Neuse River, which, fifteen years ago was one of the nation's most beautiful and ecologically abundant estuaries, is today one of the country's most polluted and distressed rivers. North Carolina has officially designated the Neuse 'nutrient sensitive' because the river is dying from the nitrogen, phosphorous, and potassium overloads. The advent of factory pork production, an industry both devised and dominated by Defendants, has contaminated that river, and many others in North Carolina's coastal plain, with a fecal and septic marinade that has killed billions of fish, sickened fishermen and swimmers, sterilized entire stretches of public waterways or choked them with fetid algae, diminished property values, and put thousands of fishermen, farmers, and outfitters out of work.

Oral argument was scheduled for two and a half months later, September 7, 2001.

As the hearing date drew near, I rehearsed answers to every question the judge could conceivably throw at me. Two days ahead, I traveled to North Carolina and checked into a hotel a few miles from the courthouse. A lawyer we'd recently hired to help with our various lawsuits, Dan Estrin, came as well. For much of the day before the hearing, he role-played judge and critiqued my responses.

The morning of the hearing, I rose early with a nervous stomach and reviewed my notes one final time. Then I carefully squeezed myself into a tailored black wool suit I hadn't worn since my law firm days when I was jogging five days a week. "I wish I'd tried this on before I left," I groaned to Dan as we met in the lobby. Now I was apprehensive, on top of everything else, about my skirt splitting at the hearing. "Don't worry. You look fine," Dan said reassuringly.

In front of the courthouse a pleasant surprise awaited us. A small group of citizen activists, including Don Webb and his wife, had gathered there to cheer us on. "Go get 'em, Nicolette," one called out. "Kick their butts!" urged another. "I will," I called back with a giggle as Dan and I headed up the courthouse steps.

Once inside the courtroom, we made our way to the front, a large wooden table at the left—the plaintiff's side. I plunked down my black nylon backpack on the trial table and unzipped it, pulling out a legal pad and pen. Filing into the right half of the room were seven lawyers wearing wool suits in various shades of gray, including a former North Carolina Supreme Court Justice. Each had a leather briefcase in hand. Our small band of supporters came in and took seats behind us. I gave them a little wave. Then I sat down, fixed my eyes straight ahead and slowly exhaled.

As makers of the motion, the herd of Smithfield lawyers went first. Their presentation was divided into three sections, with a different lawyer for each. The retired Supreme Court Justice spoke the longest, pounding his fist on the podium the most frequently and floridly embellishing his argument about why our claims, if allowed to proceed, would violate the U.S. Constitution. Nothing in their arguments took me by surprise. The judge calmly posed several questions.

Judge Malcolm J. Howard was a graying, distinguished man who'd been appointed to the federal bench by President Reagan. He treated the lawyers in his courtroom with respect. All of his questions seemed designed to clarify matters, not to stump anyone or demonstrate his own cleverness.

Then he turned to me. "Miss Hahn, you may reply." I smiled and nodded, stood and smoothed my snug skirt, then stepped over to the podium. "Thank you, your honor. May it please the Court. I am Nicolette Hahn, counsel for the plaintiffs." I heard my voice coming out strong and clear. At that moment, complete calm descended upon me. I knew their arguments, I knew the law, and I knew just what I wanted to say. More important, I believed we were right. About ten minutes into my presentation, the quiet of the courtroom was momentarily broken by someone entering the room. It did not interrupt the smooth flow of my argument.

"Your Honor, defendants portray themselves as 'farmers' exempt from federal environmental laws," I was saying. "But this contradicts the plain language of the statute, which categorizes concentrated animal feeding operations as a regulated 'point source.' A clear indication that Congress intended the Clean Water Act to govern defendant's operations can be found in Senator Robert Dole's statement about the Act. He said:

> *Animal and poultry waste, until recent years, has not been considered a major pollutant. . . . The picture has changed dramatically, however, as development of intensive livestock and poultry production on feedlots and in modern buildings has created massive concentrations of manure in small areas. The recycling capacity of the soil and plant*

cover has been surpassed. . . . [W]aste management systems are required to prevent waste generated in concentrated production areas from causing serious harm to surface and ground waters.

When I finished my argument, it was my turn to be peppered with questions. The practice sessions had prepared me well. Judge Howard's inquiries were thoughtful and interesting. He betrayed no partiality toward either side. I actually found myself immensely enjoying the back and forth.

As I fielded the judge's questions, Dan handed me a crumpled note. I opened it and saw a word of advice scrawled in Bobby's hand. He and Rick had been the ones entering the courtroom during my argument. The note said: "Remember, hog factories are a *point source* under the Clean Water Act!" It was something I had emphasized myself just before he entered the courtroom. When I retook my seat a few minutes later, the knot in my stomach finally began to loosen. Dan scribbled something on his yellow legal pad and slid it across the table. I read it and smiled at him gratefully. Written in all caps, it said: "<u>YOU</u> <u>WERE</u> <u>AWESOME</u>!"

When the hearing was over, the judge came down from the bench and shook all the lawyers' hands. The Smithfield attorneys each shook mine as well, several while flattering me about my argument. Courtrooms in the South certainly remain genteel places, at least under Judge Howard's gavel. Rick and Bobby, both pleased with the way the hearing went, came forward to congratulate me. Then I walked back to hug each of our supporters, who were beaming—Don Webb most of all. "I will never forget watching one little lady take on that whole team of big city Smithfield lawyers! Never in my life!" he exclaimed with glee.

Two television crews and a handful of other reporters were waiting outside. As I walked out into the bright afternoon sun they were already interviewing Bobby on the front steps. I felt like I'd just taken the bar exam and nailed it, a post adrenaline rush of exhaustion and exhilaration. It was a sweet moment.

Four days after the hearing, September 11, the world seemed to stop for a while. Remarkably, Judge Howard managed to keep to his schedule and on September 20, he issued his ruling: thirteen pages long and a total victory for us. He ruled that every one of our counts in both cases could proceed. A crucial line in Judge Howard's opinion stated: "The court is unpersuaded by defendants' argument that the Clean Water Act does not create a cause of action for operating without a required permit."

That sentence may not look like much, but it certainly was. The court decision was our first major accomplishment in our campaign against industrial animal operations. Every environmental agency and every animal confinement operator in the country would soon learn that a federal judge had essentially agreed that the facilities need Clean Water Act permits. The ruling really just affirmed the plain language of the statute, but the law was being universally ignored. This reaffirmation was a crucial step toward reform.

To make sure news of the ruling spread, we issued a press release. For months afterward, reporters, lawyers, citizens, and government staff were calling me, inquiring about the decision. A year after launching our campaign, I felt we were finally on our way to making some real, lasting change.

A WHISTLE STOP TOUR THROUGH FARM COUNTRY

The September day that Judge Howard handed down his ruling, Rick and I were meeting at the Champaign, Illinois, airport. We'd be criss-crossing the Midwest to make a string of speeches, which, in the spirit of old political campaigns, we were calling the "Waterkeeper Whistle Stop Tour." I got the big news about Judge Howard's decision over my cell phone as we waited by the baggage carousel. I hung up, yelped "Yeehaw!" and hugged Rick, then danced a little jig right then and there. It was a great start for our trip.

The seed for the tour had been planted several months earlier. That summer, an Iowa farmer named Wilbur Kehrli had phoned me about speaking near his northeastern Iowa town. He was a family farming advocate and liked what he was hearing from us. "I think if you'd come and speak, there'd be a lot of interest around here," Kehrli said to me on the call. Rick and I gladly accepted the invitation, always welcoming an opportunity to interact with farmers.

We were not disappointed. The gathering, held at a county fairgrounds a few weeks later, drew about a hundred people, mostly farmers and others involved in agriculture. Quite a few had heard that Bobby Kennedy was on a crusade against factory farms and their curiosity had been piqued: Was that a blessing for them or a curse? They were attentive and engaged.

Our message was that citizens, especially farmers, can and should lead animal farming away from industrialization. Rick focused on the water and air contamination from industrial hog and poultry facilities in North Carolina, playing some of his graphic pollution video footage while describing how he and his volunteer troops were countering it. "Please don't let this happen to Iowa," he implored the audience.

Then I got up. I explained what Waterkeeper was hoping to accomplish, highlighting how the Clean Water Act can be used to combat the pollution of industrial animal operations. "Every citizen, including everyone in this room," I emphasized, "has a right to file a complaint against a violator in his community." After our speeches, we took audience questions for over an hour. Some folks then stood in line as long as forty minutes to chat with us individually.

One of the people who patiently waited in that line was Jim Braun, a man in his mid-forties with graying, wavy hair and wire-rimmed granny glasses. He told us he had reluctantly given up hog farming about two years earlier. We later learned that he was one of thousands of unfortunate farmers who had been forced out of pig farming by a December 1998 market crash. After shutting down his farm, Jim became a one-man show against the dominance of big agribusiness in Iowa, advocating that the control exerted by vertically integrated meat companies was destroying not just hog farming but whole rural communities throughout the state. Working with various nonprofit organizations and individuals, Jim began spending his time trying to educate Iowa legislators and media about the issue.

Jim saw great potential in the speeches Rick and I made that summer evening in Iowa. A week later, he called me to propose that it become the prototype for a series of meetings in the Midwest. "If you guys did more speeches like that around here, I think it could really bring new energy and interest to the plight of family farmers," Jim said. I liked the suggestion right away. Rick did too. Over the next two months, we worked with Jim to get it arranged. Our itinerary had us scheduled for a whirlwind tour in late September with ten speeches in four states and additional meetings crammed in between.

When Rick and I arrived at the Champaign airport that September, Jim came to meet us. The three of us would travel together in a rented car, hitting all the spots on our Whistle Stop Tour. We'd start in Illinois, then move on to towns in Iowa, then make stops in Nebraska, South Dakota, and Minnesota. Jim had worked tirelessly to make the most of our time on this trip. For each stop, he had invited elected officials and media. Usually, we would have dinner with politicians, media, or activists before the event. Jim also enlisted the support of about fifty local organizations of diverse stripes. They represented environmentalists, farmers, people of faith, animal advocates, and rural residents. The groups' main role in the Whistle Stop Tour was to help promote the events.

As we drove around the heartland in late September 2001, American flags seemed to be displayed on every sign, building, and vehicle. But the mourning people were experiencing about the September 11 attacks did not dampen the enthusiasm of our audiences. The meetings were all well attended—ranging from 45 to 400 people. Without exception, the audiences were lively and interested. At each stop, Rick, Jim, and I would speak then open the floor for questions. Every Q&A session lasted over an hour.

Our talks focused on the water and air pollution caused by industrial animal operations. But our main message was optimistic: Traditional farming can be reclaimed by all of us. Everywhere we went, it seemed that our campaign, and particularly Bobby's leadership, was galvanizing people. It was giving them renewed hope for farming based on respect for people, animals, and the land.

Jim had also arranged for us to meet with key state officials along our route. These discussions were friendly yet formal. Rick and I used each of these meetings to further press our case that the law requires states to issue and monitor environmental permits for all industrial animal operations.

In Iowa, we met with senior staff from both the state environmental agency and the attorney general's office. The Department of Natural Resources head, Jeff Vonk, was present for the entire meeting with his staff. Director Vonk listened respectfully and posed a series of pertinent questions. At the end of the meeting he told us that he appreciated our position and essentially agreed with us. "But the reality is this," he went on to say, "we just don't get enough funding from the legislature to do these permits. We hardly have enough for what we're doing now." We thanked him for his time and his candor. It confirmed what we had already known: The laws are on the books, but politics are preventing their enforcement. Regrettably, an unfunded law is as good as no law.

Later, we followed up the Whistle Stop Tour with a letter, signed by Bobby, to the attorneys general in key farm states. It made the same point we'd repeatedly stressed on the Tour: Federal law requires animal confinement facilities to have Clean Water Act permits and there is no legal excuse for states to fail to issue them or for facilities to fail to obtain them. Bobby then made phone calls to several of the attorneys general urging them to start enforcing this law. Only time would tell what effect it would all have.

MEAT INDUSTRY MEANIES

Our Midwest Whistle Stop Tour had also included a few tussles with our pork industry opponents. On our second stop, we noticed a freckled, light haired young man in the audience whom we recognized from the first stop. Since those two meetings were more than a hundred miles apart, he was clearly no casual observer. We already knew that in response to our reform campaign, the pork industry's largest trade group had authorized more than one hundred thousand dollars for a counter attack against Waterkeeper. The money had been designated for hiring a firm known for digging up dirt and mud slinging. This towhead, we suspected, might be part of that effort.

A family farm activist traveling with us on the Tour offered to help us find out. Kendra, a pretty young blonde, sat near the mystery man and pretended to flirt, plying him with questions. He was an easy mark. Apparently wanting to impress Kendra, he bragged about himself and his weighty mission. His name was Gardner Payne (yes, that is pronounced "pain"). He'd been hired by Smithfield's lawyers to follow us everywhere on our Whistle Stop Tour, he spilled to Kendra. The law firm, he explained, wanted to nail us saying something over which it could sue us for slander.

Payne tailed us for the entire Whistle Stop Tour. To amuse ourselves, we occasionally directed remarks toward him in our speeches. But we knew we had nothing to fear from him or the law firm he worked for. Truth is an absolute defense to defamation and we were speaking nothing but the truth.

Payne wasn't the only meat industry menace we encountered. Most of them sat quietly in the back of the room, groaning or grunting at certain moments in our presentations. However, one of our detractors was much more forward. We met him in Northfield, Minnesota. At this, our final stop, Bobby joined us and gave a rousing stump speech about everything wrong with industrial animal operations to an auditorium packed with college students, farmers, and media. There we encountered a man who we felt was trying to intimidate us. His name was Trent Loos (nicknamed "Screw Loose" by community activists who knew him).

Loos' appearance, as well as his name, had a cartoonish quality. He wore a wide brimmed, black cowboy hat, three-inch stacked heeled cowboy boots and a thick, waxed handlebar moustache. He shadowed Bobby right into the bathroom just after his speech. I saw them walking out together a few minutes later. They looked like they were on the verge of coming to blows. Loos might have thought he was going in the bathroom to bully some preppy pretty boy. But Bobby is not easily intimidated by anyone and he was leaning close into Loos' face as I approached. Poking his finger on Loos' chest, Bobby was hissing at him through clenched teeth, "Back off, jerk. You don't scare me." Loos obediently stepped back several feet, then trailed at a distance into the room where Bobby would be taking reporters' questions.

Loos must have had a hard time getting over that encounter. He soon began publishing an online newsletter that seemed to be just an obsessive, running rant against us and our campaign. He called it "Truthkeepers" (a play on "Waterkeeper"), an ironic choice of names since he was later convicted of the crime of "theft by deception" in Nebraska.

Nothing done by Loos, Payne, or any of the other meat industry menaces slowed down our work. If anything, it brought more attention to our campaign and strengthened our resolve.

CHAPTER FIVE

It's a Pig's Life

THE HOG BOOM

Tracking down facts for our North Carolina lawsuits, I also unearthed the key components of pig farming's industrialization. The story starts in the 1960s when Wendell Murphy, a school teacher apparently itchy with ambition, began making money on the side by raising hogs. At the time, there were slightly more than a million pigs in North Carolina, a dozen or two on every family farm, things looking very much as they had since the early twentieth century. North Carolina was not a major pork producer. But in 1962 Murphy would ignite a chain of events that would soon bring about a hog population boom.

That year he decided to construct a feed mill on his parents' property in the tiny town of Rose Hill, thereby closely following the trail blazed by poultry pioneer and fellow southerner Jesse Dixon Jewell. Murphy soon offered his neighbors a per-head fee to do the hog raising for him. Like Jewell, Murphy provided farmers the animals and the feed, while farmers kept the manure and the full legal responsibility for properly disposing of it. Soon Murphy was pioneering the use of a housing method, modeled on poultry facilities, in which pigs were kept confined twenty-four hours a day. Corporate contract swine production was off and running.

As Wendell Murphy's business burgeoned, he lifted his sights beyond southeastern North Carolina. In 1983, he successfully ran for the state assembly and headed off to Raleigh. Over the next ten years, Murphy served in the legislature while continuing to expand his hog contracting business. As a member (eventually vice-chairman) of both the agriculture and appropriations committees, and former college chum of the governor, Murphy oversaw the passage of a series of laws benefiting his business. The statutes shielded industrial animal facilities from nuisance lawsuits and (purportedly) from the need to comply with various environmental laws. By the mid-1990s, Murphy had built the largest hog empire in the nation, grossing more than $600 million in 1997.

But even the seemingly indomitable Wendell Murphy was headed for tougher hoeing. The expansion that Murphy helped trigger led to a national glut of hogs and complete market meltdown in late-1998. Prices to farmers plummeted to 8 cents per pound, meaning farmers were being paid far less for pigs than the 35 cents per pound cost of raising them. Adjusted for inflation, hog prices were lower than they had been during the Great Depression. Farmers simply had nowhere to go with their hogs. Some were giving them away. Some desperate confinement operators were rumored to be tossing live newborn pigs into their manure pits to avoid the cost of raising them.

The market collapse forced thousands of independent hog farmers throughout the country, including our Iowa friend Jim Braun, out of the business for good. Murphy was not immune from the fallout and saw his company's bottom line soften. Shortly thereafter, he began serious talks about a merger with Virginia-based

Smithfield Foods, the country's largest hog slaughterers. For Smithfield, the hog market crash was actually a windfall as it could buy all the pigs it wanted for almost nothing while the price consumers paid for pork didn't go down. On January 30, 2000, Smithfield announced it had completed acquisition of Murphy Farms, Inc.

Although Wendell Murphy reportedly disliked the equally colorful man seated at the top of Smithfield, Joe Luter (yes, that's pronounced "looter"), the marriage of the two companies is said to have made Murphy a billionaire. A year later, Murphy rightfully accepted his place in North Carolina's Pork Hall of Fame.

Like the big poultry processors, Smithfield's Luter had been longing to expand the reach of his domain from slaughterhouses into animal raising—previously the exclusive province of farmers. Luter considered it essential to creating absolute uniformity in his pork products. By acquiring Murphy's operations and several other companies, Smithfield gained control over every aspect of pork production "from piglets to pork chops," as is said in the business.

Even before combining with Murphy, Smithfield had been making its mark on North Carolina. In 1992, under the name Carolina Food Processors, Smithfield had opened the world's largest hog slaughterplant in Tar Heel, North Carolina, killing more than thirty thousand pigs every day. When the killplant had been on the drawing board, citizens worried about its environmental and community effects spoke up in opposition. The state's Division of Environmental Management's director had reviewed the project and advised that the nearby Cape Fear River was already dangerously contaminated with nutrients. The river could not take the additional pollution that would come from the slaughterhouse, he declared. In response, Smithfield promised to create four thousand jobs. Soon, the governor personally intervened and granted the company a waiver from the mandatory environmental impact statement.

The combined effect of the activities of Murphy and Smithfield was to swiftly squash virtually every surviving glimmer of traditional hog farming in North Carolina. When Murphy first built his feed mill in the early 1960s, more than seventy-five thousand family hog farmers in North Carolina were raising about a million hogs a year. By the day we held our Raleigh press conference in 2001, the number of pigs raised in the state had skyrocketed to ten million, but it was on fewer than three thousand farms. Virtually all of those were confinement operations, three-quarters of which were owned or controlled by Smithfield Foods.

North Carolina had become the China of pig production. Non-existent environmental enforcement; low-priced land; low-wage, nonunionized labor; subsidized grain; and mechanized methods all combined to enable it to put out artificially cheap pork. To compete, Iowa and other states were compelled to match North Carolina's low cost industrial methods. By the time we launched our campaign, virtually all pig farming in North Carolina, and most of it in the United States, had converted to corporate controlled confinement production. Remaking pig farming into an enterprise that was kinder to animals and gentler on the environment would be no small task.

INSIDE AN INDUSTRIAL HOG OPERATION

About two weeks after I had mailed our legal notice letters to hog operations, I received an intriguing phone call. The voice on the other end of the line identified itself as one of the letter recipients. He was a "contract grower" for Smithfield Foods. "Ya'll have made a mistake here," he urged in a thick, Carolina drawl. "I run a clean operation." I told the man that these were serious charges, there'd been no mistake, and he needed to get himself a lawyer right away. "I'd like ya'll to come on down to my place and take a look around. We're not causin' any pollution here," he persisted. The guy seemed sincere in his belief, at least in an O. J. Simpson kind of way, that his facility was not a source of pollution. I felt a pang of sympathy for him. But just a pang; I was too well acquainted with his state file, which was chock full of odor complaints and manure lagoon spills. "Sir, you really need to get an attorney," I advised him firmly. "We'd like to come see your operation, but I'm not comfortable doing that until you're represented by counsel."

That call was my first invitation to penetrate the bowels of an industrial hog operation, which, until then, I'd only peered at from public roadways and in photographs. In the years following, I would tour the insides of more than a dozen, including the facilities of that contract grower. Some were operations we were threatening to sue. Others belonged to farmers who just wanted to show us that confinement operations weren't as bad as we were out there claiming.

Rick and I cherished the chance to experience the facilities firsthand so we always consented to these offers. But our acceptances were draped in dread. Even before stepping foot in them we knew we'd be entering dank, putrid buildings crammed with thousands of suffering creatures. After one particular day of touring pig confinement buildings in North Carolina, Rick turned to me and said, "Nicolette, this has been one of the most depressing days of my life."

Pig operations are segmented. Some facilities keep breeding sows for piglet production; these facilities are farrowing operations. (In pig farming "farrow" means "give birth.") When the piglets are two to three weeks old, they are taken from their mothers and moved to another part of the facility, referred to as a "nursery," which keeps them about five weeks. From there, pigs are shipped to a "growing operation" and finally, a "finishing operation," where they are fattened in the final weeks leading up to slaughter. From finishing operations, pigs are trucked to slaughterhouses and killed at about five months of age.

Throughout its operations, agribusiness maintains a singular obsessive focus on formulating consistent, lean pork. Carcass uniformity is treated as a paramount virtue. To achieve it, agribusiness owned or contracted facilities raise only animals that come from what is essentially the company store—central operations that produce breeding females or vials of sperm. By design, then, as in the poultry industry, there is no longer any real genetic variation in industrial pig herds.

And the pigs are being bred lean in the extreme. A *Saveur* magazine article I saw once described pigs in industrial operations as resembling Dachshunds. Much of that is due to zealous selective breeding. Additionally, the drug ractopamine is now being routinely administered at industrial hog facilities (Paylean is a common brand). The drug acts like a steroid, building muscle at an unnaturally rapid rate. Although the substance is currently legal in the United States, its use is banned in many other countries. In July 2007, contamination by ractopamine residues was the basis for China's rejection of large shipments of U.S. pork.

The industry justifies such breeding and doping for extreme leanness by pointing to consumer demand for lower fat foods. There may be some truth to that. But the real driver is a more direct economic motivation. Pound for pound, muscle is much more valuable than lard, especially these days. Integrated meat companies make less profit on every fatty pig.

Not surprisingly, the extremely lean hog is problematic both for the pork eater and the animal. The consumer ends up with dry, flavorless meat. Famed cookbook author and *New York Times* food writer Mark Bittman has described industrial pork as "turning to sawdust" when cooked. Many chefs and retailers I've met attribute recent decades of flat pork sales to the poor eating quality of modern industrial pork. For the pig, the problem is quite a bit more serious. That's because the pig's internal temperature, both keeping cool and staying warm, is controlled by its body fat. When the animal is engineered by humans for excessive leanness it loses its ability to regulate its body temperature in response to fluctuating external temperatures and changing seasons. Most modern industrial hogs would, quite literally, perish outdoors.

As one can easily imagine, a pig living in an industrial facility is also constantly stressed by its contaminated, intensely crowded surroundings. Twenty-four hours of every day the animal breathes and bathes in air laden with ammonia, hydrogen sulfide, dust, viruses, bacteria, and endotoxins. This takes a heavy toll on the pig's immune defenses, which are engaged in a never-ending battle just to keep the animal alive. For this reason, most pigs raised in industrial confinement are dosed daily with drugs in their feed or water. One USDA report estimated that 97 percent of hogs in finishing operations are continuously given antibiotics in their daily food or water.

Staying healthy under these virtually unlivable conditions is made still more challenging because the animals are denied access to fresh air, sunshine, and exercise, which all bolster a body's defenses. My parents, like good parents the world over, I'm sure, were constantly nudging us: "Go outside and play!" It was plain common sense that to be healthy, kids need plenty of time outdoors, exercising and breathing fresh air.

This was once accepted wisdom among those practicing good pig husbandry as well. The highly regarded book *Swine Management* by Iowa State University animal husbandry professor Arthur Anderson (published in 1950), repeatedly emphasizes the importance of exercise to the raising of healthy pigs. Here's a typical passage:

> *Encourage sufficient exercise. Exercise plays an important part in the development of strong bones, strong muscles, with resultant healthy pigs. Plenty of exercise, therefore, should be encouraged, preferably on pasture, or at any rate on a dirt run. Avoid confinement to concrete or other floors that keep the pigs from the soil.*

Equally ardent about the necessity of exercise was *The Hog Book*, the 1913 quintessential American hog farming manual by H. C. Dawson. Throughout the book, the author extols the virtues of porcine exercise. In one passage, for example, he urges that during pregnancy, "it is essential for the sow to take plenty of exercise daily."

However, the pigs that Rick and I saw in industrial facilities never got exercise, fresh air, or sunshine. Three decades earlier, agribusiness had abandoned as quaintly old fashioned the common sense notion that all living creatures need these basics. We toured pig facilities of each life-stage. All were heavily mechanized, with automated feed, water, and manure handling systems. Each contained thousands of animals and, with one exception, was a mile or more, sometimes many miles, away from the residence of the person in charge. Some facilities were left entirely unattended about twenty-three hours of every day.

One visit to a farrowing operation was particularly memorable. "Well, ya'll would have to take *showers* before going in there," the facility owner had sternly warned us when we expressed interest in seeing the breeding part of his operation. He seemed hopeful that this information would suffice to discourage us. I did not relish the thought of going through the facility's "bio-security" gauntlet but nothing was about to deter me from witnessing for myself how the sows live, the most criticized aspect of industrial pig raising. "Oh, that's no problem," I assured him cheerily in response.

With that, our guide ushered Rick and me, one at a time, to the rear end of a confinement building. I entered a makeshift locker room that reminded me of a camp changing area. As I dutifully showered and donned the head-to-toe white paper spacesuit and plastic boots provided me, I had the unsettling sense that someone might be watching. All of this, we'd been told, was to prevent us from transmitting any illnesses to the pigs.

Scrubbing up like a surgeon to visit some farm animals struck me as quite surreal. All the more so because although we were being made to act like we were entering a sterile hospital ward, I already knew that it would be more like entering a sordid medieval dungeon. The incongruity of the cleansing ritual struck me with full force as I stepped from the showering room into the smelly, dimly lit, manure-caked interior of the confinement building. But I knew that like a patient in the operating theater is susceptible to infection, these pigs were in fact highly vulnerable to disease.

The breeding facility we were touring that day was divided into several sections. In the first, scores of pregnant females were kept in individual metal containments, referred to in the industry as "gestation crates." My paper spacesuit swished as I walked up and down the aisle looking over the caged pigs. Arranged in long rows, each crate held a 300 to 700 pound animal and was less than two feet wide and five feet long. Gestation crates are designed to virtually immobilize the sow, so each crate was intentionally too narrow for her to turn around. She could not lie down without bumping up against the metal bars. Scabs resembling bedsores and abrasions were scattered everywhere on the herd's sides, stomachs, and legs.

I'd often read that pigs are as smart and curious as dogs so I was getting a sick feeling as I looked at these intelligent creatures living like this. Even a backwater dog pound gives each animal space to move about and comfortably lie down. It occurred to me that maybe I could offer one pig a tiny olive branch from humankind. I paused and knelt down in front of a sow who'd been watching me approach. "Hey girl," I called to her softly. Eyeing me with suspicion, she instantly pulled back. Pressing herself against the back end of her cage, she seemed to loathe being singled out. My gesture was too little and much too late.

I straightened up and asked the grower why the sows were kept in cages. "Oh, it keeps 'em from fightin' and guarantees they each git enough feed," he explained. "So, you're saying the sows are in crates for their own benefit?" "That's right!" he cried, apparently delighted that I had comprehended what others had difficulty grasping. "To tell the truth, I don' think they mind being in the crates at all," he added a moment later. "They git used to it real quick." Pointing to several of the animals, I asked why the sows were chewing on the bars of their cages and waving their heads from side to side. "Oh that's perfectly normal," he assured us with no hesitation. Rick and I exchanged a glance.

Our guide then asked us if we wanted to see how the sows were impregnated. "We want to see it all," Rick responded dryly. "Great. Follow me," our guide directed. We then went to an adjoining building filled with still more rows of sows in metal crates. There, we stood to one side as a large adult male, a boar, was led down the alleyway between the cages. "You can tell which ones are in heat by the way they act around the boar," one of the workers explained.

The workers then turned their attention to a sow who had just signaled that she was ovulating by groaning, pricking her ears and raising her tail. Into the sow's swollen red vulva a worker swiftly stuck one end of a four-foot plastic tube. It looked like the half-inch clear plastic tubing found in any hardware store. Holding high the tube's other end, another worker then poured a couple tablespoons of white liquid from a small bottle. We were told it was company-provided sperm. "There. Now she should be pregnant soon," our guide narrated with some satisfaction. The boar and sow never even came in contact. So, that was it. Sex, industrial agribusiness style.

The inseminated female would stay in that restrictive metal crate for her entire pregnancy of three months, three weeks, and three days. Just before giving birth, she would be moved to another, slightly wider, metal crate. That crate would allow the mother to lie on her side so the piglets could nurse. These "farrowing crates" also severely restrict a sow's movements. The piglets are in the crate, too, but are separated from the sow by metal bars. Other than providing swollen teats for milk, the mother can barely interact with her piglets.

After witnessing the insemination, Rick and I were led to the section with farrowing crates. Here were the sows just about to give birth and those with newly born litters. We paused to admire a wriggling pile of pink piglets clambering over one another to hungrily nurse on their crated mother. Rick and I knew that these piglets would be weaned in a couple of weeks and the grim cycle would begin anew. Within a few days after the piglets' weaning, the mother sow would be re-impregnated with the little bottle and the plastic tube and then placed into the gestation crate for the duration of her next pregnancy. And we knew that this grinding regimen would likely burn out this sow after just a few years (about one-fifth her natural life span) at which point she would be sent off to the slaughterhouse to become the main ingredient of breakfast sausage or salami.

Despite that dreary prospect, looking at these piglets before me I could not keep from smiling. I was taken with how exquisitely formed they were, so different from human infants, who barely resemble adults. "They're perfect little miniatures," I whispered to Rick. "So beautiful." "Do you want to hold one?" our host offered. I nodded. He reached over the side of the crate and picked up one of the two-day old piglets, which seemed to go completely unnoticed by the sow. I cradled the tiny warm creature in my arms, pressing my lips against his fuzzy pink forehead as he quietly consented to the attention. I put him back only reluctantly. I suspected it would be the sole taste of human kindness this pig would ever know.

Our tour was almost over and Rick and I stopped to gaze at a few more litters on the way out. "Watch your step," Rick said with chagrin as he pointed to a lifeless piglet, just larger than the one I held in my arms, lying across the passageway. This one they apparently had forgotten to clear away before we walked in.

THE WILLY WONKA POOP FACTORY

Another unforgettable tour was of hog operations belonging to a man named Chuck Stokes. With thirty confinement buildings, he claimed to be the second largest contract grower in North Carolina and was experimenting with new methods of waste management. It was one of my regular visits to the state for hog related research. Don Webb, Rick, and I met at the Webb home in the morning and drove to the Stokes property together in Don's SUV. As we approached our destination near the town of Ayden, Rick pointed to a grand white nineteenth-century farmhouse with a well-maintained yard and garden. "There's Chuck's house," he said. We continued on about a mile to a cluster of four metal buildings with a swampy three-acre manure lagoon out back. A dozen or so black cattle were grazing not far off.

We pulled up near the enclave of buildings and parked next to the only other car on the site. After walking around a few minutes, we found the car's owner, a man in his sixties who introduced himself as Don Lloyd. He was not there to take care of the animals. His single focus was their millions of gallons of liquefied waste.

Lloyd had come to orient us to his handiwork, an intricate Willy Wonka-type contraption for handling this hog facility's urine and feces. This was his pilot project, his baby, which he was hoping to one day replicate and sell by the thousands. As soon as we all shook hands, Lloyd launched into an animated sales pitch. "My system," he began dramatically, "will one day totally eliminate the need for liquid manure lagoons!"

Lloyd kept talking as he led our trio into an outbuilding the size of a large garage, which was filled by his device. Enthusiastically, he pointed out three two-thousand gallon tanks for waste storage and a spaghetti of pipes and tubes to move the pig sewage from points A to B and beyond. A detailed explanation of the contraption's workings ensued. It boiled down to this: pig urine, feces, and flush water go in one end and are separated by the time they reach the other end, with solids then going one way and liquid going back to the buildings for flushing. With an additional device (not present here), the solids wastes could, in theory, be dried and pelletized, he explained. Then Lloyd drew our attention to the computer screens monitoring the entire system.

The poop separating contraption looked pretty pricey and energy consumptive, so I raised the question of cost. Lloyd said he planned to sell the devices for about $150,000 and that it takes at least three pig confinement buildings (more than 3,000 hogs) to make the numbers crunch. Then he revealed what he considered the best part. "What's really great is that this whole system is designed so that *no one* needs to be here at all!"

To me, however, getting humans even further removed from

the care of the thousands of animals enclosed here did not seem like a very good idea. In fact, the whole thing looked like a lousy idea. I know that from a certain perspective, technology like Lloyd's sounds sensible. He and other inventors claim they can rid the world of manure lagoons, thereby eliminating major threats to waters and air. And given the many environmental and human health threats posed by lagoons, that proposition has undeniable appeal.

But such systems are actually compounding the problems of an approach to farming that is fundamentally flawed. Industrial animal operations have taken the simple act of raising animals for food, which humans have been doing in relative harmony with nature for some ten thousand years, and have made it technology-dependent, polluting, and harsh. "Techno-fixes" like Lloyd's further complicate a simple process while often causing new, unforeseen problems. They're also expensive and tend to depend on public funding. Worst of all, such devices drive agriculture even further toward large-scale, concentrated animal raising. They don't pay for themselves unless done on a big scale, sometimes a very big scale. So, to Lloyd's Willy Wonka Poop Factory, I was thinking, "No, thanks."

A CONTRACT GROWER WITH SOME REGRETS

Just as Don Lloyd was winding down his spiel, Chuck Stokes arrived and apologized for being late. He was a heavy set forty-something man in a red polo shirt with a crew cut and an intense gaze. Stokes was known in North Carolina as a hog operator with his own brand of vigilante flavored political activism. He co-founded the group Frontline Farmers whose sole discernable purpose has been fending off and attacking critics of the hog confinement industry. Among other activities, the group has been known to pressure universities whose research documents the negative effects of confinement production. Several members of Frontline Farmers had attended our

North Carolina Hog Summit—(actually, *we caught two attempting to sneak in without paying!*)—and actively sought out the reporters to provide quotes contradicting our speakers.

Given his history, Rick, Don, and I were somewhat wary of Stokes' intentions in inviting us. But for the moment, Stokes seemed to want to take a more conciliatory approach. We welcomed the chance to tour his farm and didn't mind listening to his rap.

Chuck spent a lot of our visit reciting his frustrations about being a confinement operator. He finds himself in a bind, he said. Years ago, he had borrowed hundreds of thousands of dollars to put up confinement buildings, dig pits for lagoons, and purchase equipment for manure spraying. But he now knows from firsthand experience that this system doesn't really work. "I sure bought into the lagoon and sprayfield system," he told us. "But it was a mistake." Like every other confinement operator, he's had plenty of troubles with pollution and odor. His problem is this: What does a person do after he has invested everything he has—and more—in a system that he now knows to be a failure but is still years away from paying off?

How does he like being a contract grower, I asked. "Well, I appreciate it," he responded readily. "It has sustained me and my family." Indeed, it seemed to be sustaining them quite well. Nonetheless, Stokes had no kind words for Smithfield. He expressed resentment about the power of the large agribusiness corporation to which he is entirely beholden.

Stokes also claimed his view of industry critics has evolved from his earlier days. "When all these environmentalists started coming after me, I thought they were a bunch of psychopaths," he said as we walked toward one of his manure spreading fields. "But after a while, I realized they were saying some things that made sense." And now, he said, he's weary of fighting them and troubled about having to constantly defend himself from a barrage of complaints by his own neighbors. "I'm tired of being the bad guy in my hometown."

We couldn't be sure of Chuck Stokes' real reasons for wanting to talk to us. Maybe he hoped we'd somehow end up helping his business. My guess is that it was some combination of motives that included a genuine shift in his view of manure lagoons, based on years of unpleasant firsthand experiences.

In addition to working with Don Lloyd's pilot project, Stokes was also trying out a new method of spreading manure on fields as an alternative to aerial spraying. As we walked over to observe the system in action, Stokes explained this second part of his experiment with poop. When we reached the field's edge, we all stopped and stood side by side. It was a plowed field planted with some sort of grass. Stokes pointed to a tractor forty yards from us dragging what looked like a giant tail. The tail was perforated and had brown liquid spurting from it like jets of a fountain. "This way, I'm putting the manure right on the field without the spraying. It really cuts down the odor." It was true that the odor wasn't bad from where we were standing.

However, the tractor application system was another big expense and it was unclear whether it would yield any concrete benefits. Stokes told us he paid around $140,000 for the tractor and thousands more for the application equipment. Even if it cuts back the odor, which it appears it may, I'm skeptical that this method can decrease pollution. The manure is still untreated and the amount of manure being applied to the land isn't being reduced. What isn't taken up by the plants or bound to soils will still end up in the water or air.

Regardless of their effectiveness, I don't see Chuck Stokes' various experiments with poop as the answer to the industrial farming problem. None of these technological marvels does anything to address the abysmal daily conditions for workers and animals inside confinement buildings. And because of their price tags, they are generally out of reach for smaller operators. Whether or not the various technological approaches will work, at least Chuck Stokes was trying to fix a system he had come to admit is deeply flawed. I had to give him credit for that.

A few feet away from where we stood watching the tractor spreading pig manure, a second group of beef cattle were grazing the field. Some were standing in raw hog waste up to the tops of their hooves, forced to eat vegetation splattered with brown slop. The scene gave a whole new meaning to the phrase "grass fed beef."

Our last stop at Stokes' place was some of his confinement buildings about a half mile beyond the manure application field where we'd been standing. As we got out of our cars, I recognized the familiar rotting egg odor of industrial hog production. Chuck led the way up a ramp into the first building. As we entered, the pigs simultaneously retreated to the far corners of their pens like schools of frightened fish in a tank. It is an experience I've had every time I've walked into a pig confinement facility. This building, a "growing facility," held juveniles, about three months old, which should be a wonderfully curious, playful age. But here, not one pig was at play.

The second building, a "finishing facility," held larger, older pigs close to slaughter age. "I hope you're not too thin-skinned," Chuck remarked with a self-conscious chortle as we walked in. He was referring to the carcass of a dead hog heaped near the entry door. I'd seen dead pigs on every facility tour. "Well, I'm used to it," I replied with a sigh.

In both buildings, all of the pigs' tails were cut off to stubs. This practice of tail docking is universal at confinement pig operations. In every industrial hog facility I've toured, I've seen nothing but truncated tails. The tails are generally clipped off with wire cutters—and without anesthetic.

Like dogs, pigs are said to communicate with their tails. They wag them when they are contented, let them drop straight down when sick, and stick them straight out when frightened or alarmed. Yet the U.S. pork industry claims that tail docking is necessary. It says that pigs bite each other's tails and that the tails can then become infected. When pigs' tails are cut off, the stubs stay intensely sore and so, the theory goes, the bite will cause so much pain that the bitee will move away from the biter. (The industry refers to this as "avoidance behavior.")

Part of this is true: Tail biting is common among pigs in confinement. This I know both anecdotally and from the research. But tail biting is really a direct result of how they're being reared—in metal buildings with concrete floors, giving pigs nothing to occupy their active minds. In nature, pigs spend most hours of their days rooting around in the dirt, exploring and grazing. Stuck inside, bored pigs often bite one anothers' tails—one of the many so-called "vices," or abnormal behaviors, that occur when pigs are raised in confinement.

Research does not support the industry's claim that tail docking reduces tail biting. In fact, several European studies show that cutting off pigs' tails has little or none of the desired effect. A 2003 British study, for example, actually found the reverse, concluding: "tail docking was associated with a three-fold increase in the risk of tail biting." The Animal Welfare Institute prohibits tail docking in its pig farming protocols. The standards state: "A behaviorally appropriate environment and good nutrition normally eliminate the need for routine tail-docking."

Because its benefits are dubious and it causes animals to suffer, docking pigs' tails has been prohibited in the European Union since 1991. However, neither the U.S. Department of Agriculture nor the domestic pork industry has shown the slightest inclination to follow suit.

After finishing our tour of Chuck Stokes' confinement buildings we told him we appreciated his frankness and his time. As Rick, Don, and I drove off, I felt relief to be putting some distance between ourselves and that place. After discussing what we'd just experienced, we intentionally turned our conversation to more pleasant subjects and considered whether to stop for something to eat. "I don' think we'd betta," Don growled with a twinkle in his eye. "Truth is, we *reek* of hog crap!" He was right. We had all the windows down to air ourselves out but the smell was clinging to our hair and clothing. We decided to pass on dinner. Anyway, I'd sort of lost my appetite. On the drive back to Don's house he pointed out dozens more hog and chicken confinement operations along the way. The drama we'd spent the day witnessing was being repeated at every one.

A WOMAN ALONE
FIGHTING FOR THE ANIMALS

The day after going through the Stokes operation (and wearing a clean set of clothes) I headed west, to Asheville. It was a pre-arranged pilgrimage to meet Gail Eisnitz, someone I'd long admired. She is the author of *Slaughterhouse,* a detailed book that documents widespread problems in the treatment of animals at U.S. slaughterhouses. In her research, she'd interviewed more than one hundred workers, including dozens of Smithfield employees. I'd read her book a year earlier and wanted to talk directly to Gail about her many experiences wrestling with the meat industry.

From her book and by reputation, I knew her as a determined, courageous woman who spent years doing groundbreaking work for animal protection organizations. Gail is famous among animal activists for her daring—even dangerous—undercover investigations. In one, she went dumpster diving at hog operations, cutting eyeballs out of discarded pig carcasses to test for the illegal drug clenbutorol (which her testing did confirm).

Before my visit, I'd heard that Gail wasn't well but I didn't know any of the specifics. To avoid imposing, I had promised her a brief visit. When an attractive fortyish blonde answered the door, I was pleasantly surprised to find Gail smiling and appearing perfectly healthy. "Come on in," she said warmly. We immediately fell into the comfortable conversation of old friends. Soon we decided to carry on in a downtown Asheville café.

Over lunch, I prompted Gail to review her years of animal advocacy for me. I learned that much of it had been grueling work done entirely alone. She had especially worked on getting the press to cover animal cruelty, something she'd often found an uphill battle. "Most of the time, the media just refuses to run these stories because they say their viewers or readers will be turned off. They're businesses and they don't want to lose people," she explained with exasperation. Because her efforts had mostly been behind the scenes, Gail rarely received public recognition for her accomplishments. When I pointed that out, she replied simply "Well, I've always felt it was more important to get the work done than to get the credit." And I could tell she meant it. I admired her genuine humility.

At the same time, I also detected that Gail felt proud of her accomplishments. Other than the book, she seemed most pleased at having provided the idea and background information for a major story that ended up on the front page of the *Washington Post* in 2001. The article described a Washington state cattle slaughterhouse where workers recounted that every day dozens of animals were being dismembered while still fully conscious. "They die piece by piece," a slaughterhouse laborer said of the cattle he was butchering every day on the line. The worker had spoken to Gail because he was so disturbed by the situation. I remembered the article well. "That story," Gail told me as we finished our meal, "got more public response than any other *Washington Post* article ever—except one they'd run during Watergate."

The *Washington Post* piece was especially notable for having kicked off a chain of important events. Senator Robert Byrd, a great lover of animals, was so outraged by the story that he read parts of it aloud on the Senate floor. Gail learned of this and searched for an effective way to communicate with the Senator, hoping to encourage him to take further action. She decided the best way would be with the help of a renowned animal expert. "Senator Byrd didn't know me from Eve, but I thought if someone of Jane Goodall's prominence contacted him, he'd take her calls." Gail's hunch was right. With Goodall as intermediary, Gail provided Senator Byrd additional information and ideas for legislative action. A short while later, Byrd introduced a Senate resolution to enforce the

Humane Slaughter Act. It passed by wide margins in both houses of Congress. "Senator Byrd still doesn't know who I am, as far as I know," Gail said with a sly smile that seemed tinged by just a touch of regret.

After lunch, Gail and I strolled to a nearby coffee shop as stories of our work lives continued pouring out of us. "I'm so glad you're here. I've felt so isolated for such a long time," Gail confided over her coffee mug. "It's a relief just to have someone to talk to—someone who really understands." For decades, Gail felt up to doing risky, gruesome investigations, uncovering, documenting, and exposing horrific acts of animal cruelty. Somehow, she had always been able to keep herself just enough removed. But now she found herself wondering whether all the awful things she'd seen over the years were coming back to haunt her. "I kept thinking I had seen the worst that humans could do to animals, and then I would see something even worse," she said, slowly shaking her head.

Of all the nightmarish scenes she witnessed, the rows upon rows of caged sows now disturbed her the most. "Where is God for all of those sows?" she asked me in a shaky voice. "I've always believed there's a reason for everything. But what could possibly be the reason for millions of sows spending their entire lives trapped in those crates?" I've always shared Gail's faith that things happen for some purpose and was looking for something comforting to say, but I couldn't think of a good reply. Instead I just nodded and said, "I know. I know."

At the time of my visit, Gail's two most recent investigations were of Midwestern hog raising facilities. At both facilities, she spent months getting to know local people, including current and former employees. Her likeability and sincerity clearly aided in gaining their trust. At both operations, the hired workers themselves had become deeply troubled by conditions at the facilities. So much so that they began talking to her, documenting the problems on their own, and eventually even bringing her inside the facilities, allowing her to photograph and videotape what was going on.

For one large hog raising facility in South Dakota, Gail recorded the testimony of twenty-one past and present employees. Their words graphically describe sick and injured pigs left unattended to die, whole herds drowning in their own manure from backed up flush systems, and frequent pig cannibalism. Before my visit I had read excerpts from their testimonies and that of others in a powerful petition Gail had filed with the South Dakota attorney general requesting him to prosecute the company for animal cruelty. He was refusing to take any action despite her compelling case.

I complimented Gail on the South Dakota petition. "Thanks. But I'm disappointed that the attorney general's not prosecuting, and about how little press attention it's all gotten," she responded. The press comment surprised me. With Gail's assistance, the *New York Times* did a full-page story with photos, painting an unflattering portrait of the facility. How could she hope for more than that? "Oh, *that*. That was *so disappointing*," she replied with a dismissive wave. "I got the reporter out there, even got her inside the operation. She saw everything." Okay. So far, so good. "But then the *Times* gave the operation three days official notice before it came out to take pictures and do interviews! You can do a *lot* of clean up in three days. It makes me wonder what's happened to the whole idea of investigative journalism." I could see her point.

My visit with Gail, meant to be short, had now been going on for six hours. And I felt like we were just getting started. Gail seemed to share my sentiments. She invited me to stay overnight, and I accepted. Back at Gail's apartment, she suggested we look at some of her latest video footage. I wasn't savoring the prospect, but somehow it seemed wrong to refuse. My sense was that Gail needed to share them with someone. Then she could verbalize her distress and perhaps even exorcise some demons in the process. The tapes were of the South Dakota hog operation and one in Nebraska that Gail was also investigating.

She put in the Nebraska tape first. It showed a sow facility, a dimly lit chamber with hundreds of adult females in narrow rust-ing metal cages. It had a much grungier look than any facility I'd toured, and the crowding was extreme. The automated feeding system was apparently broken, so a worker was feeding manually. He was pushing a metal cart down an aisle between the cages and simply flinging scoops of feed at the pigs. The alleyway was so narrow the cart kept banging against and getting stuck between the cages. It seemed the facility's owner wanted a caged sow on every possible square inch.

The sound coming from the tape was worse than anything I'd ever experienced. Apparently reacting to the feed cart, the hungry sows were all loudly shrieking. It must have been the norm at this place—the worker was wearing air traffic controller headphones.

With the remote in hand, Gail fast-forwarded through parts of the tape, pausing now and again to point things out. "See all those abscesses?" and "God. Just look how skinny those sows are—their bones are sticking out. They're practically starving to death!" She stopped the tape again to show me baby piglets up to their necks in liquid manure, flailing their legs to stay afloat. They had fallen through a rusted hole in the floor and were now trapped beneath the metal grate. This part of the tape had been filmed by two of the workers. On the audio, they agonized over what to do. The workers couldn't figure out how to rescue the drowning piglets, who were now far from the hole they'd fallen through. I could only imagine the distress of the mother pig.

Gail's gory videos were difficult to watch yet hard to turn off. After we'd been glued to the screen for about an hour she put in one of several tapes from the South Dakota operation. The first video showed numerous skittish and sickly animals. I spotted more than a dozen with soccer-ball or baseball size protrusions from their stomachs, sides, or necks. These were not the sick-pens; they were just ordinary areas of the daily operation.

Then Gail put in a second tape from South Dakota with a dis-gusting sequence of clips showing pigs eating one another. "Once a sick one goes down," Gail narrated in a trance, her eyes fixed on the screen, "the others begin to nibble on it and eventually it turns into a downright feeding frenzy, even though the pig's still alive." At this part, I just had to look away.

After what felt like a long time, mercifully, Gail turned off the VCR. I think that by that point, we'd both seen enough. "All these years, I've been able to do this work, and I've been okay. But for some reason, these tapes have really been getting to me," Gail said with her eyes downcast. "Especially seeing all those caged sows— waving their heads, chewing the air. . . . It's like an insane asylum!" I had no trouble understanding why these pictures had been plaguing her.

Lying on Gail's sofa in the quiet dark that night, the pictures and the sounds of the caged sows ran through my head, too. I slept only fitfully and repeatedly woke to thoughts of those awful images and noises. I kept wondering how people buying pork from industrial facilities would feel if they glimpsed these videos, even for just a few moments. Anyone would have to feel moved. "This must be hell on earth," Gail had remarked.

The next morning, as I was getting ready to leave, Gail handed me some documents, saying, "I want you to have a copy of these." One was a report released earlier that month by the federal government's General Accounting Office. The GAO had audited slaughterhouses for their compliance with the Humane Slaughter Act and found that the law was being routinely and flagrantly violated throughout the meat industry. Inhumane slaughter conditions were the norm. The report's findings were an enormous vindication for Gail's work.

Gail walked me to my car. Before I climbed in, I thanked her for her warm welcome. I hugged her tightly then grabbed her shoulders and looked her squarely in the eyes. "Gail, your work is so important. Please don't ever forget that." My visit was only a brief respite from her isolation. As I pulled away, I waved to Gail thinking about how now, once again, she would be on her own.

Back on the road toward Raleigh, mountains enveloped me. I hummed along to bluegrass music coming over the car radio. Half a dozen black cattle grazed peacefully on a hillside. Seeing them triggered a realization that after driving around the state for a week I had encountered almost no farm animals on the land, only handfuls of cattle sprinkled here and there. It was a startling thought because North Carolina is not only the country's second largest hog producer, it's also the largest turkey producer and fifth largest chicken producer. The animals are all there, but they're living inside buildings, out of sight. Only the intrepid investigator and occasional covert video expose their plight to the outside world. The animals are easily forgotten.

The next time I saw Gail was six months later in Washington, D.C. The Animal Welfare Institute was recognizing her with its highest honor, which it grants only once every few years, the Albert Schweitzer Award. Her spirits were high and her health seemed good.

A DIFFERENT KIND OF PIG FARMER

My work against industrial animal operations was taking me to rural communities all over the country for research, meetings, and speeches. Along the way I met dozens of hog farmers. A couple of them, I came to know especially well. One was Jim Braun, the erstwhile hog farmer and guide of our Midwestern Whistle Stop Tour. His experience is typical of conventional hog farmers in the latter part of the twentieth century.

Jim had been reared on a diversified pig farm in Latimer, a north-central Iowa town. When he went off to college he deliberately avoided studying agriculture, certain that he'd "never, *ever* raise hogs again," as he told me. After graduation he tried his hand at several professions, even living briefly as a missionary. But his family badly needed him home. So after a while, he was drawn back, becoming his family's fourth generation of Iowa hog farmers. Soon after, he helped radically alter the way his family farmed. With what struck me as a blend of pride and embarrassment, he told me, "My father and I were among the first farmers to embrace confinement technology for raising hogs."

Thousands of families like Jim's created an Iowa historical fabric tightly interwoven with pigs. Jim's great-grandfather had founded the farm after emigrating from Germany. He was among the many European settlers who brought hog farming know-how and a devotion to supping on pork, helping establish the state as America's largest hog producer. It's a distinction Iowa has held for a century, today raising more than 15 million hogs annually.

Historically, Iowans embraced this hog heritage, respectfully referring to hogs as "the mortgage lifters." Conditions were and remain ideal for raising pigs because Iowa has some of the world's highest quality soils and ample moisture, both essential for growing the feed crops of corn and soybean. (Like humans, pigs are omnivores: they devour grass when on pasture, but unlike cattle, cannot survive on it.) Hogs reliably multiplied the value of grain crops, substantially boosting a farm's cash flow at times when crop prices were low, making it possible to pay the mortgage in the lean years.

In spite of Jim's help and its conversion to confinement technology, the Braun farm struggled to stay afloat. With each passing year, he and his father felt more tightly squeezed by integrated pork companies. Their independence seemed to be slipping through their fingers as "the integrators" (as Jim and other farmers call big meat companies), gained control of Iowa's pork industry. Eventually things got so bad that Jim was forced to close down the farm. The Braun hog buildings have sat idle ever since.

That sad demise to the tradition of a hog farming family is typical of U.S. farmers. Around the time that Jim had returned to farming, there were almost a million farms with hogs in the United States. When Jim shuttered the Braun farm, less than three decades later, there were fewer than one hundred thousand.

Fortunately, Jim's experience, while widespread, is not universal. A man named Paul Willis, another pig farmer I befriended, has a very different story. The early part of Paul's life was a lot like Jim's—growing up with hogs on a central Iowa farm then going off to college confident that he would never return to the farm. With no thought of pursuing agriculture, Paul majored in psychology then spent three years in Africa with the Peace Corps. When he returned to the United States, Paul lived for several years in Wisconsin and Minnesota recruiting for the international corps and working for the domestic peace corps in the VISTA program. Like Jim, Paul one day made up his mind to go home to Iowa and help run the family farm. To this point, their life stories run parallel.

However, soon after, Paul's course veers away decisively. When he started farming again Paul knew he wanted to raise hogs. It was the early 1970s and everyone he was talking to was saying that the future of animal agriculture would be in confinement facilities like those pioneered by the poultry industry. Anyone who didn't accept the inevitability of that cold, hard fact was clinging nostalgically to an obsolete past. But Paul has always marched to his own drumbeat. His gut told him he didn't want to keep pigs that way.

A lot about confinement hog raising bugged him. For one thing, there was the farmer's life. Instead of working outdoors with fresh breezes blowing on your cheeks and honking geese migrating overhead, those farmers were stuck inside ugly metal buildings. And he'd heard they spent much of their time dealing with manure handling problems.

Then there was the way the animals lived. Paul had always liked pigs. Since boyhood he had found spending time with the animals the best part of farm life. He liked watching them being born, growing, experimenting, and learning—just enjoying their lives. When he returned to farming he wanted to see piglets playing in pastures and sacked out in the sun, sows grazing on grass and building nests for their young. He wanted to live alongside pigs foraging the fields, as his father had done and farmers had been doing for thousands of years.

Paul was a speaker at our North Carolina Hog Summit, but amidst the day's whirl of activity he and I barely had a chance to talk. Our first real conversation occurred in Minnesota, on the final stage of our Whistle Stop Tour. A stocky, sandy-haired man with a handsome, weathered face and bright blue eyes was wending his way in my direction through a packed room. Upon reaching me he pressed a Petite Peppered Ham into my hands, saying with ceremony, "This is to thank you for everything you're doing." With equal gravity, I thanked him for the ham (of course, I left out mentioning my dietary habits and later shipped it to my parents who were delighted to receive it). I also told Paul how pleased I was to finally have a chance to thank *him*. He was invaluable to our campaign by actually doing in Iowa fields what we were advocating from our New York offices. He was farming in a way that was "no longer possible," according to an endless flood of agribusiness propaganda.

In the years following those first encounters, I've passed many enjoyable hours in the company of Paul Willis and his wife Phyllis, cherishing my time on their farm. The Willis family farm, which closely resembles the one I had tenderly guarded in my imagination until witnessing industrial production, rekindled my faith in the possibility of raising pigs in a way that respects nature and animals. There's a red wooden barn, fields of gently swaying corn and soy, a prolific vegetable patch, cats scurrying around haystacks, pigs roaming in meadows, and colorful chickens scratching in the yard. The Willis farm is also home to a sizeable native pothole prairie, lovingly and laboriously restored over many years by Paul and his family, returning that landscape to how it looked before the arrival of European settlers.

When Paul first came back to his family's farm, he set about starting his own pig herd. By a stroke of good fortune, a neighbor soon called and offered to sell him a sow and her litter. The sow had what was already considered at that time to be "old fashioned" breeding: plenty of fat and hardiness to thrive outdoors. Paul jumped at the chance, and that sow became the foundation of his herd. He's always taken great pride in that first sow and has carefully selected every breeding animal since. Like every good traditional farmer and rancher I've met, much of the pleasure Paul takes in farming has come from working to improve the genetics of every generation on his farm. He carefully selects the right females to keep as breeding stock and seeks the best possible boars to mate with them. Mating has always been done naturally, by simply putting a sow or gilt (a young female) into the same field with a boar at the appropriate time.

With these time-honored methods, Paul gradually grew the size of his herd. An intellectually vibrant man, Paul has eagerly sought the latest research on the care and feeding of pigs. But he was never interested in putting them in metal buildings or funneling their manure into cesspools. His animals have always lived on pasture, eaten a drug-free daily ration, and have never spent time in metal crates. Paul grew his own feed corn and soy, rotating his pigs and crops yearly on various fields, which benefited from the nutrients in the pig manure. Paul's decision to stick with the traditional methods of raising pigs has allowed him to keep his costs low and be profitable. He never had to take out the large loans needed by confinement operators for capital-intensive structures.

Over the years, as Paul labored on his farm, confinement hog production was going up all around him. He could always sell his family's pigs but he received no premium for the way he was raising them. His hogs ended up at Hormel, just like the ones from confinement buildings down the road.

That would change once he met Bill Niman in the mid-1990s. Neither could have known what a fateful event their meeting would be. The idea had originated from a Peace Corps pal of Paul's, Jeannie McCormack. After her service in the Peace Corps she, too, had returned to her family's farm, a sheep ranch in the hills near Rio Vista, California. Bill Niman was selling the McCormack family's lamb to fine restaurants in the Bay Area under the name "Niman Ranch," a fledgling company supplied entirely by a small group of traditional farmers and ranchers. The company and the farmer network were spearheaded by Bill Niman. He had started raising pigs and cattle just north of San Francisco in the early 1970s and by this point was the Bay Area's most respected meat purveyor.

Bill and Paul have slightly different recollections of their original encounter, but it went something like this. Paul was in California to visit both his sister and Jeannie McCormack, who arranged for Paul and Bill to meet. The two men hit it off instantly and decided to keep talking. Bill recalls being intrigued by this self-assured farmer proclaiming the superiority of his pork. Bill wondered: How much better could it be? He would soon find out.

Upon returning to Iowa, Paul headed straight to his freezer and grabbed a couple of chops for express mail to California. When Bill received them he immediately threw one into a sizzling pan and pulled out a knife and fork. Biting into the meat, he understood why Paul was so confident about the virtues of his pork. The flavor amazed him. Like what he'd eaten as a kid, only better. "It was the best porkchop I'd ever tasted," Bill fondly reminisces.

Bill soon realized that he and Paul shared much more than a fondness for a good piece of meat. He and Paul began a regular dialogue about how they might be able to collaborate. They discovered they were the same age and shared many life experiences and core beliefs. They often spoke about farming, especially the way they raised animals. The men had a common understanding of meat as a special food that should only be produced with the utmost care. Soon Bill took a road trip to Paul's farm in Thornton, Iowa, and he liked what he saw.

In February 1995, Paul sent Bill the maiden load of pork from Iowa on a refrigerated truck. When Bill delivered the meat to some of the Bay Area's best restaurants, including the culinary landmarks Chez Panisse and Zuni Café, chefs responded with untamed enthusiasm. Many chefs were seeking meat that was free of the chemicals and drugs common on industrial operations. But what thrilled them more was the way Paul's pork ate. A lot of them had a reaction like Bill's: it was simply the best pork they'd ever had.

The demand for the Willis' free-range pork was so strong that

Bill soon asked Paul if he had any neighbors raising pigs the same way. To create a year-round steady supply, Paul began recruiting other like-minded farmers to join them. The advantages for the farmers were tangible. For one thing, they were paid a premium price. Bill believed that farmers raising animals a special way should be rewarded.

On top of that strictly economic benefit, there was tremendous satisfaction for the farmers in knowing where their meat was ending up. When farmers sold their pigs to Hormel or Cargill, the pigs were taken away in a truck and the farmer never heard another word about it. The meat they were producing ended up in the vast stream of anonymous, undifferentiated commodity pork. For farmers selling to Bill Niman, however, it was totally different. They knew the stores and restaurants where their meat was offered, prominently designated on labels and menus as Niman Ranch. Moreover, the organization performed regular taste-testing of all the meat, providing farmers direct, specific feedback about the quality of their pork. Stores and restaurants were proud to be serving it. Farmers felt honored to have the products of their farms promoted and to know whom they were feeding.

Around the time Paul and Bill began working together, Paul also approached Diane Halverson of the Animal Welfare Institute. She says (although he denies it) that he came up to her in the parking lot after a community meeting and presented her with a ham. (Based on my personal experience with Paul's calling card, the Petite Peppered Ham, I am inclined to trust Diane's recollection.) What's certain is that Paul wanted Diane to endorse his farm.

"Everyone in the meat industry was scared to death of animal advocacy groups, but I didn't feel that way at all," Paul has told me about why he sought out Diane. "I felt that if I was doing something wrong on my farm—I wanted to *know* it." Diane told Paul she would be happy to tour his farm and advise him on how he could improve the lives of his animals.

When Diane did visit Paul's farm, she found little to criticize.

"I was impressed with the way the animals lived and the attitude toward them, which was one of respect, and even affection," she recalls. Both Diane and I believe that much of this ethos emanates from the loving spirit of Phyllis Willis. She is a tall, strong woman whose arms generously dole out warm hugs to almost everyone who crosses their threshold, a classic farmwife in the very best sense of the word. Phyllis is quick with a smile, a kind word and a deep, heartfelt laugh. And she never seems to feel too busy to stop and help a person or animal in need.

One of my favorite Phyllis Willis stories is when one day she drove by a field and noticed a sow and her piglets in the wrong place, outside of the area where they should have been in. Somehow, they had all slipped through the electric fencing. The sow and her babies were in no immediate danger, but Phyllis worried about them being separated from the herd and having no access to the water troughs and feeders. She promptly found Paul on the farm and told him about the situation. "Okay, I'll take care of it," Paul assured her. Later that day Phyllis happened to pass that field again and noticed that the pigs were still there. She again went and found Paul, reminding him about the wandering pigs. Paul responded that he had his hands full but that he'd get to it soon enough. Phyllis was not satisfied by this answer. She decided she'd better try to take care of it on her own.

"I knew I could never chase down those piglets," Phyllis told me, "so I figured I had to try to think the way they do. I knew I couldn't force those pigs back into that field so I had to make them *want* to go in," she continued. "So what I did was I tacked up that electric fence and just crawled under the fence on all fours. And you know what happened? The pigs followed me—*every one!*" When Phyllis relayed the tale I could tell the memory of it still delighted her.

Just before I first had the pleasure of first visiting the Willis' idyllic setup, I had been immersed for months in industrial animal agriculture. It was therefore a tremendous relief for me to see the way the people and animals live together on their farm. Actually, I felt pure joy. Paul and I walked together through several meadows, which were dotted with small huts occupied by clusters of mothers and their young. Like cape buffalo on the African savannah, sows were moving slowly through the vegetation, grazing while surrounded by their piglet clans. For a long time, I stood ankle deep in the green growth, watching the piglets running, rolling, and happily chasing each other through the field.

Taking in the scene on Paul's pastures I was happy to see full tails waving in the air like pink ribbons. Paul follows the AWI husbandry standards, and anyway he's found that he does not need to dock his animals' tails. His pigs have plenty to occupy themselves. And, he told me, "I like seeing pigs with their tails."

I asked Paul if I could hold a piglet. His sows were friendly, alert, and very much alive. They were also attentive mothers, making my request somewhat challenging. "Well, let's see if we find one who's bedded down with her pigs," Paul suggested. We quietly snuck up behind one of the huts and peeked in. The mother was lying on her side on a thick bed of straw nursing a litter of about eight little ones. Paul stuck his arm in and retrieved a perfect little spotted pig in his palm. Seconds later, the sow's open mouth lunged after Paul's arm like a crocodile leaping after its prey. She was not about to take the theft of one her babies lying down. "Yikes! Never mind," I said. We hastily relinquished her kidnapped piglet.

I knew then that the pigs on the Willis farm had their wits about them and their instincts intact. In fact, they were living a lot like their cousins, the wild boars. The farm seemed guided by a single principle: Let the pigs live in a way that respects a pig's nature. When they have each other, good feed, fresh air, sunshine, and ample exercise, the rest generally takes care of itself.

On one of my many visits I sat with Phyllis and Paul around their Formica kitchen table sipping coffee and, as usual, enjoying a lively conversation. Phyllis was holding an insulated mug. Beneath its clear plastic surface was a picture of a lovely smiling pig with wings and a halo. "I like that," I remarked casually, assuming it was mass produced Hallmark merchandise. "Ah, well that is a very special picture," responded Phyllis, "because that was a very special pig." The mug, she explained, had been made by her daughter Sarah to honor a favorite resident of the Willis farm named Trixie.

Trixie was one of three sisters in a litter. The others were Dixie and Pixie. Each of them, Phyllis told me, had special personalities—unusually interested in people and very smart. "I used to feed them apples and they got very tame," Phyllis reminisced. "Every evening when I checked the water they'd come bounding over, chewing my shoes and rubbing up against me. They always wanted their ears scratched. It got so I felt I couldn't go in there without bringing them a treat."

Paul and Phyllis can talk for hours about the animals they've known over the years. Every animal is an individual. They both innately believe that each animal deserves a good life and, of course, a swift and painless death. "The animals we raise have only one bad day," Paul says simply.

When I met Paul he and a small staff had built a network of several hundred pig farmers raising pigs like he does. As I write this, the organization has grown to a co-operative of almost six hundred farms. All of the pig farms follow the Animal Welfare Institute standards and send their livestock to Niman Ranch. The farmers have the satisfaction of knowing where their meat is going and they are getting a premium price for practicing traditional farming. Everyone from Smithfield to Tyson/IBP is now trying to imitate (at least in a superficial way) their success. The farmers of Niman Ranch give me great hope for the future of pig farming.

CHAPTER SIX

A Door Closed, a Door Opened

LEAVING A JOB I LOVED

About twenty months after moving to New York, I took a trip to Michigan. My parents, siblings and their spouses were gathering to celebrate Christmas at my childhood home. The snow fell steadily, thickly blanketing the ground in Kalamazoo. A group of us braved the icy back roads to the Wahmhoff farm for the ritual felling of a blue spruce. Back at the house, we built a big fire and decorated the tree. The days following were filled with cookie baking, carol singing, present exchanging, and church going. Every day I cross-country skied in my father's favorite field. I was utterly relaxed and happy in this blissful retreat. It was the first time I'd felt that way in months.

Upon returning to New York, I was preparing to go in to the office when I realized that a black cloud was forming above me. I was filled with trepidation. A few months earlier, the fledgling Waterkeeper board had finally secured a professional executive director after an exhaustive national search. This was excellent news, I thought, until my first meeting with the woman they'd hired. My new boss told me she had already determined that the campaign against industrial animal operations, of which I was in charge, was "far too large" and consumed "way too much of the budget." "But the campaign is flourishing—it actually *brings* money to the organization," I explained. Well, anyway it was too removed from our core mission, she continued, and had to be scaled back drastically. And another thing, I would have to work on other issues. "In fact, I've decided you shouldn't be spending more than half your time on this project," she announced with stunning finality. I was floored. "But there's so much work. This project could use a staff of ten!" I protested. She appeared unmoved.

For a while, I was hopeful that with a bit more time on the job the new director would come around to seeing the campaign's value. But that was not to be. Instead, she became increasingly insistent that the project be downsized and that I drop many of the initiatives I'd gotten off the ground. Soon she turned critical of my work, often in front of others.

Shortly before going home for Christmas break, I'd seen a recent photo of myself and realized with horror that my thick, long hair was noticeably thinning. Dark circles hung beneath my eyes. A good night's sleep had eluded me for months and I looked ten years older than when I was hired. The position had always been stressful and demanding, but that hadn't mattered to me when I felt I was affecting positive change and my work was valued. Now, I was slaving away but was reporting to someone who I felt was hostile toward me and to my work. The job I once loved had turned into a nightmare.

That first day back from vacation, I made the decision to quit. It was simply a matter of self-preservation. I asked for a meeting with Bobby, who remained an ardent supporter of the reform campaign and me personally. "I can't stay here any longer," I told him. He was aware of the trouble I was having with the new director. "I don't blame you, Nicolette. But I wish there was something I could say to make you stay." There wasn't. Although I hated the thought of leaving, my mind was made up.

Then I met with the executive director and informed her I'd be departing four months hence. (She could barely refrain from springing from her chair and clicking her heels in celebration.) I knew that Dan Estrin would admirably carry forward the litigation. But I didn't want any part of the reform campaign to suffer from my departure and worried about the rest of it. To smooth the transition, I would help find and train my replacement; I'd make sure all of the files were in order; and I'd complete the second Hog Summit, which I'd already begun planning and was scheduled for March. Then I'd go.

Where I would go, I had no idea. Somehow I needed to find a way to keep advocating on the issues that had become the very purpose of my life these past two years. Perhaps I'd set up a new nonprofit organization dedicated to ridding the world of industrialized animal production. Maybe I would write a book.

But for the moment, my main focus would be organizing the second Hog Summit, which would take place at the Surf Ballroom in Clearlake, Iowa. That would be my last official act in this job, so I wanted it to be even better than the first. As in the preceding year, the program would be designed to demonstrate the feasibility of environmentally sustainable, humane farming, not just delve into the ills of industrial animal methods. That emphasis was especially important because this year's event would take place in the heart of America's farmland. Good turn out and media coverage would again be essential.

Rick and I, along with Diane Halverson of AWI, set to work. Calling heavily on the wide network of sympathetic experts and activists we'd built together in the preceding year and a half, most of our time was spent trying to draw speakers, attendees, and media. We especially wanted good participation from farmers, whom we considered absolutely essential to the movement's ultimate success.

Among those who agreed to speak were cattle rancher Bill Niman and hog farmer Paul Willis. Both of them enthusiastically addressed our Iowa audience about how they farmed and the organization they'd spent years putting together. Bill's talk was imbued with an evident passion for raising animals using traditional methods, which benefit farmers, animals, the land, and people eating the meat. He stressed the strong and growing demand for natural meat among consumers and food professionals. "Chefs and food lovers around the country want meat with great flavor that only traditional farmers can really provide. And they want it from farms raising their animals outdoors, without drugs and chemicals." I enjoyed listening to Bill speak and admired his passion. But my personal interaction with him that busy day was limited to little more than a perfunctory exchange of pleasantries.

The summit turned out to be a resounding success. Almost a thousand people trekked there from twenty-four states. The speeches were informative, heartfelt, and inspiring. The audience heard from scientists, economists, citizen activists, and animal welfare experts who gave data and testimonials that were both damning to industrial production and hopeful about alternatives. As in the preceding year, we nourished the gathering with meals featuring Niman Ranch pork.

The day was topped off once again with Bobby's inspiring invective. Don Webb first revved up the crowd with some of his classic fire and brimstone. Bobby then took the stage and launched into a denouncement of agribusiness magnates, whom he compared to the robber barons of the late nineteenth century. "Agribusiness is invading our rural communities, driving independent farmers out of business, stripping towns of local control and capturing state legislatures with its political clout. This is threatening our democracy!" he railed. "I consider this an even greater threat to our country than foreign terrorists because we will always unite against and ultimately defeat any foreign enemy. But the enemy within—the one that slowly and almost imperceptibly erodes our freedoms— that's the most insidious threat." The crowd ate it up.

Before stepping down, Bobby acknowledged a string of people in the room, including the new executive director. The audience politely applauded after each introduction. He saved his last words for me. Bobby spoke for several minutes about my deep commitment to changing the way farm animals are raised in this country, how much I loved this work, and the many people I'd tried to help over the past two years. "She's been at her desk until eleven at night for months to put this event together." Then he asked me to stand. From my seat in the third row I stood and turned to wave at the crowd, which included hundreds of people I'd met over the preceding two years. To my amazement, the audience rose to its feet and gushed rowdy cheering and applause. It seemed to last a very long time. My eyes brimmed with tears of gratitude. I could not have asked for a more meaningful send off.

VEGETARIAN AND CATTLE RANCHER UNITE

"You never know who's doing you a favor," a friend of mine used to say. Never have those words seemed truer than regarding the Waterkeeper boss who made my life hell. Soon it became evident that I walked away at just the right moment.

A Saturday morning, about two weeks after my last day on the job, my cell phone rang. It was around 8:30 and I was still in bed. Not recognizing the number, I answered with grogginess and hesitation. "Hello?" "Nicolette? It's Bill Niman. I'd like to thank you for all the ways you've been helping Niman Ranch by inviting you to dinner tonight." A food writer for the *New York Times* would be joining us, too, he hastily added. Sounded legitimate. Still clueless about my next career move, I welcomed the networking opportunity. "Sure. That sounds great. Where and what time?"

I had first met Bill Niman a year earlier, at an awards ceremony in Washington, D.C., hosted by Animal Welfare Institute.

On behalf of AWI, Bobby was presenting the Albert Schweitzer Award to the Polish Farmers' Union president, Andrez Lepper, for his work preserving family farms in Poland. Of course, by that time I was already favorably disposed toward Bill Niman. For months, we'd been holding him up publicly as a rare golden child in a meat industry riddled with bad boys.

Bill Niman is a tall, handsome man with dark bedroomy eyes and perfect posture, something I've always found inexplicably alluring. But his pleasing attributes escaped me at this first encounter. My clearest thoughts about Bill that day were that his dark wavy hair desperately needed to be tamed with a haircut and that he was in urgent need of a pair of pants that were not blue jeans. And then there was his moustache. That definitely had to go. Needless to say, it was not love at first sight, at least not for me.

But I did like the man. There was an appealing modesty and sincerity about him. When I complimented him by saying that he was a true agricultural pioneer, he muttered, "Well, I'm not doing anything differently than lots of other farmers and ranchers." And when I asked him if he wanted to be introduced to Bobby he said he'd like to meet him, but only if it wasn't too much trouble. It was a refreshing switch from the mobs usually tripping over me to shake Bobby's hand.

Bill, for his part, later confessed that he felt "struck by a thunderbolt" at that first encounter. Years later, he still remembers exactly how I had my hair and what I was wearing, right down to my shoes. But he is a patient man. A very patient man. He waited over a year to make his move. That was when he called that Saturday morning to invite me to dinner in Manhattan. And though it was largely unwitting, he chose just the right moment. Had I still been in the throes of my Waterkeeper job, I could hardly have broken away for dinner, let alone a serious courtship.

But anyway, that first dinner was a business meeting, as far as I was concerned. I mean, aside from the issue of his moustache, which was serious, I could not possibly be interested in a cattle rancher. True, most of the men I'd dated were meat eaters, and I'd never opposed other people eating meat, but how could a long-time vegetarian get involved with someone who produced meat for a living? The idea was absurd.

Dinner was at the Savoy, a charming, intimate restaurant in Soho that meticulously sources its ingredients. I ordered a pasta primavera and Bill took note. "Are you a vegetarian?" "Well, actually, I am." "Oh, that's cool," he replied without skipping a beat. Clearly, this was no ordinary cattle rancher. After dessert, the journalist left but Bill and I lingered. It surprised me to realize how thoroughly I was enjoying myself. I had not smiled or laughed so much in a long time.

When we finally got up from the table, we walked out together and I hailed a cab. Bill leaned into the front and handed a twenty dollar bill to the driver, one of the few American-born cabbies I've ever had in New York. Pulling away from the curb the driver turned to me and, in a heavy Bronx accent, said, "He's a real gentleman. I don't see dat anymore. *You outta marry dat guy*!" I chuckled at the silly suggestion.

Over the next few months, Bill and I saw each other quite often. I was living off my savings and researching for a book I'd decided to write about farming's industrialization. Whenever Bill was in New York, which was fairly frequently, we'd take long walks in Central Park, get together for coffee in the afternoon, or share a dinner at a restaurant that served Niman Ranch meat. I even found myself developing a fondness for Bill's moustache. Still, I kept him at arm's length. At twenty-two years my senior, I considered him too old for me. And, of course, there remained the matter of his business. While I admired and appreciated his approach to raising animals, I struggled to imagine a romantic involvement with someone whose very livelihood was producing meat.

By contrast, my last boyfriend had been a youthful, clean-shaven, Ivy League educated, hardcore vegan. On paper he seemed perfect. He eschewed all meat, fish, dairy, and eggs, and refused to wear leather, wool, and silk. No honey allowed, either. And, as it turned out, he wasn't terribly sweet. Ironically, the experience actually opened my mind to the meat man I was now getting to know. My close-up view of the vegan approach to life had made me realize it was not for me. I especially disliked the self-righteous attitude adopted by many of the vegans I encountered.

I also witnessed how lots of vegans subsist on utterly unnatural foods. My boyfriend's cupboards and freezer were stuffed full with highly processed, industrial creations. His meals were built around pretend-meats like soy (chicken-flavored) "mcnuggets," soy (chicken-flavored) "buffalo wings," and pink and white plastic looking (pork-flavored) fake "bacon." There was even a tofu "jerky" imported from Taiwan. Not only did I find this stuff unpalatable, it was totally artificial.

Something about Bill Niman, however, was wholly real. In spite of our differing diets, our world views were remarkably in sync. The more I came to know him the more clearly I could see that Niman Ranch's unique environmental and humanitarian values originated from Bill. And the more I could see that he was a thoughtful man of great integrity and sensitivity. He also defied almost every stereotype of people in the meat industry (exemplified by statements he's prone to making, like, "People should eat less meat"). In his three decades plus of ranching, Bill Niman had become a modern day Good King Wenceslas of sustainable agriculture—arduously forging a path that others can see and more easily follow.

About three months after the cabby bestowed me with advice to marry this man, Bill and I were sitting at a small table by a Central Park pond enjoying the tiny boats' meanderings. It was a warm, sunny afternoon and we were sipping champagne from plastic flutes. Bill was wearing a black shirt.

Suddenly I was the one struck by a thunderbolt. Why hadn't I previously noticed that this man I respected and admired was also pretty damned attractive? *Very* damned attractive, actually. And I loved him. There was really nothing complicated about it. Very soon after that, we decided to get married. (I should note that there's some disagreement about who proposed to whom.)

A year later, on a joyous day in my hometown, surrounded by family and friends, we exchanged vows. By our wedding day, my health and fitness had been fully restored along with a thick, full head of hair. Nights, I was sleeping like a baby.

Soon after, I moved from Manhattan to the first and original Niman Ranch, Bill's home in Bolinas, California, just north of San Francisco. A chapter in my life that started desolate had ended very well, indeed. Somehow I had miraculously gone from frazzled urban lawyer in a small apartment with no serious love prospects to happily married woman on a seaside cattle ranch. Within months of the move, I'd be spending half my time on horseback, working as a rancher. None of it would ever have happened had I remained chained to my Waterkeeper desk. Maybe one of these days I'll send my old boss a thank you card.

CHAPTER SEVEN

Beef, the Most (Unfairly) Maligned of Meats

CATTLE RAISING— MUCH AS IT EVER WAS

Arriving at my new home in California, I was a ranching neophyte. My contact with the beef industry while at Waterkeeper had been minimal. Occasionally, a citizen would call me troubled about a cattle operation. However, compared to the deluge of complaints pouring in about poultry, pig, and dairy facilities, such calls were few and far between. Because we were geared toward countering industrial animal raising and the beef cattle sector has undergone comparatively minimal industrialization, little of my time ended up focused on beef.

Knowing Bill's marrow-deep commitment to ethical animal stewardship, I had imagined I would be comfortable with what took place at his ranch. I already knew that the animals were well treated, he used no chemicals or hormones, and he never allowed anything other than grass, hay, and grain to be fed to the animals. But I had also suspected I would be as discouraged by mainstream beef production as I'd been by industrial poultry and pork. That turned out not to be the case. Not by a long shot, really.

To my surprise, much of what I've come to understand about how beef cattle are raised in the United States has actually been heartening, especially when compared with other food animal sectors. By learning the ins and outs of our own ranch, visiting dozens of other ranches and numerous cattle feedlots, and by reading extensively about cattle history and husbandry, I discovered that most American beef cattle are raised in a way that I find morally acceptable (with the exception of certain feedlot practices, which I'll describe momentarily). The overall cattle husbandry method is similar to how humans have been tending cattle for thousands of years.

My bovine education included lessons in zoology and history. Cattle belong to the zoological order *Artiodactyla* (Greek for "even-toed"), suborder *Ruminantia* (cud-chewing animals that are all strictly herbivorous). Most modern cattle are believed to be descendants of a much larger, now extinct beast called the aurochs. The book *Animals that Changed the World, The Story of the Domestication of Wild Animals* calls cattle domestication "the most important step ever taken by man in exploitation of the animal world," but notes that how and when it happened is largely unsettled. Recent research suggests that bovine domestication may have begun as far back as eleven thousand years ago, independently in two or more locations, probably India and the ancient Near East.

Cattle first arrived in the Americas in 1493, with Columbus' second voyage, which brought Spanish breeds, mostly as draft animals. But not until 1591 did the original seed stock for Latin America's cattle herds arrive, when a Spanish merchant named Gregorio de Villalobos shipped a small group of Andalusian breed cattle to the New World. European settlers of North America later landed with early British and Continental breeds, primarily intended for labor and milk. As settlers pushed south then pioneers headed west, cattle were part of the migration, proliferating particularly in areas with abundant native grasslands. Many of the earliest cattle ranches were in the South, especially in Georgia and the Carolinas. "The long grazing season, mild winters, and extensive, sparsely wooded uplands were especially favorable for beef production," notes one history. "It was said that a steer could be raised as cheaply as a hen."

By the mid-1800s, many cattle ranches were being established on the grasslands of the west. But the bulk of cattle raising was taking place in the heart of the country. "Prior to the Civil War, the Ohio and upper Mississippi Valley states constituted the center of the beef cattle industry. On practically every farm of this area was a herd of beef cows."

Contrary to what's often suggested in the popular press these days, neither ancient European nor early American cattle husbandry was strictly grass-based. According to *A Short History of Farming in Britain*, as early as the Middle Ages cattle were fed *exclusively* "hay and corn" (corn used in the British sense of grain) during the winters. Moreover, in the newly forming United States, cattle grazed on grass from spring to fall, then ate hay and grain over the winter. Likewise, mature steers were fattened on hay and grain before slaughter. Describing the Ohio and Mississippi Valley regions in the mid-nineteenth century, a history of beef cattle explains that calves were kept until three or four years old, when they were put into feedlots and fattened on corn. "Grass was still abundant and relatively cheap and constituted the sole feed during the summer and fall; while hay, either timothy or prairie, supplemented with a liberal allowance of shock corn, formed the common winter ration."

By the 1870s, it was realized that beef cattle breeding herds could most economically be kept on the open ranges of the western states, whereas fattening of cattle was most economical in the Midwest Corn Belt. In contrast to the Midwest, the Far West offered vast grasslands but lacked the rich topsoil and reliable rainfall necessary for significant cultivation of most feed crops. It was no longer profitable to maintain sizeable breeding herds in the Midwest, since Midwestern feedlots could be filled with western steers at a much lower cost. "As these western cattle proved highly satisfactory for feeding purposes, the breeding herds [of the Midwest] were rapidly dispersed, and the beef cattle of the Corn Belt became largely a finishing proposition."

By 1880, 50 million people lived in the United States and 27 million cattle dotted the country—from Maine to California, from Florida to Washington state. Although they were fed hay and grain at certain life-stages and times of year, the animals' ubiquity was largely due to their ability to survive for most of the year on a simple diet of grass, making them easy and economical to keep.

The fundamentals of beef cattle raising have changed little since the early days of America's settlement. Unlike chickens, turkeys, pigs, and—as I'd soon learn—dairy cattle, cattle raised for meat still spend their lives outdoors, much of the time on spacious pastures, grasslands, and the arid open ranges of the Far West. They are selected for hardiness and survivability on the range, traits that serve any animal well, whether domesticated or wild. Females kept for breeding are chosen for even temperament, good mothering tendencies and physical traits that will enable unassisted birthing. The choice of animals selected for breeding is made by the family that owns the ranch (and owns the cattle), not some distant corporation. Breeding is usually done by simply putting bulls and mature females into the same pasture at the right time of year, and letting nature take its course. Beef cattle not only are fully capable of courtship and natural mating, they seem to immensely enjoy it.

Mother cows typically spend their entire lives on pasture. They have one offspring per year and raise each of their young until an appropriate weaning age (generally between seven and nine months). Calves are nurtured by the milk and constant teachings of their attentive mothers, supplemented by ample doses of guidance from the entire village of mother cows. With few exceptions, cattle ranches keep their animals as a herd in a physical environment similar to the one for which they evolved. In short, beef cattle live much as nature intended with surprisingly little human interference.

Likewise, cattle ranchers spend many of their waking hours outdoors, functioning a lot like traditional cattle herders. More than anything else (other than fixing fences), their time is spent just observing the animals—making sure none is sick or injured, none is separated from the herd, and every one is in the right place. By necessity, people who raise cattle are closely attuned to every whisper of change in the seasons and weather. Knowledge of rain and snowfall, temperatures, wind, and sunshine is vital to their success. These variables determine the grazing conditions and, consequently, the health and nutritional needs of their animals at any given moment.

There's plenty to criticize about current beef industry practices, which I'll talk about in a bit. Nonetheless, even the worst aspects of modern cattle raising are less troubling than the daily norm at industrial poultry, hog, and dairy operations.

That will probably surprise a certain segment of people, who think of beef as the most problematic meat sector. I've heard several people attempt to establish their social consciousness by exclaiming something like, "Well, of course, I *never* eat beef!" (perhaps between forkfuls of industrial chicken). I suspect negative impressions like these have been stoked by popular books and articles from journalists like Jeremy Rifkin, and recently Michael Pollan, who have trained an especially intense spotlight on beef, probably because it is the most expensive and cherished meat in the Western world. In the United States, beef has been commonly referred to as the "King of Meats" and dethroning any royalty makes for good copy. Such writings create the impression cattle ranching is the most suspect of all animal farming.

But I heartily object. First and foremost because cattle raised for meat generally lead good lives, making beef the most humanely raised meat. A friend of mine, a vegan who works for the national Humane Society, agrees. "People who care about animals are better off eating a steak than an omelet, if the eggs come from factory farms with battery cages," he's often argued. Of course, the lives of beef cattle aren't perfect; there will be moments of stress, discomfort, and pain. But such moments will be part of every animal's life, including every human's. And in stark contrast to the miserable daily existences of animals in industrial facilities, many beef cattle can truly enjoy most of their days, living alongside their herdmates and other species of animals, almost as beasts in the wild. Certainly, life in nature can be, as Thomas Hobbes famously opined, nasty, brutish, and short. Nonetheless, I think any of us would choose it over an existence mired in suffering imposed by humans for the sake of convenience and economy.

Secondly, the people involved in cattle raising can lead very good lives. I've witnessed their immense satisfaction and pride in their occupation. And there's good reason for it. Their days are not passed toiling in depressing metal buildings crowded with stressed and sickly animals. Instead, their work is largely outdoors, tied tightly to seasons and weather, inhaling fresh air and interacting with thriving creatures whose bloodlines they have carefully developed over generations. This all engenders in cattle ranchers a healthy admiration for their animals and an intense and meaningful connection with the natural world.

THE INDISPENSABLE GRAZING ANIMALS

Contrary to the impression created by exposés about the beef industry, responsible cattle raising can be environmentally benign, even beneficial. It's true that the earth has experienced much overgrazing, including by cattle. But recent decades of scientific research are dispelling the myth that cattle grazing is *intrinsically* damaging to the environment. For example, a 1998 report for the World Bank and United Nations states:

> Conventional wisdom suggests that much of the blame for 'desertification' and land degradation in arid rangelands rests with pastoral

livestock production. There is now a considerable literature which corrects this misconception on two counts: the extent of dryland degradation is greatly exaggerated because underlying ecological dynamics have been misunderstood, and the contributory role of livestock has been misspecified.

Real world examples and empirical studies have shown that good herd management can prevent the erosion and land degradation associated with overgrazing. Carefully managed grazing can even foster rather than diminish vegetation. Properly timed grazing triggers beneficial biological processes that improve soil and actually stimulate the vegetative reproduction of grasses "to produce an average of 30 percent to 45 percent greater herbage biomass," a North Dakota State University study determined.

Cattle and other grazing animals coevolved with grasslands, making them essential to maintaining grasslands. In North America, large ruminant mammals covered much of the continent prior to European colonization. This included huge populations of deer, antelope, an estimated 10 million elk, and 30 to 75 million bison. In 1806 the journals of Lewis and Clark noted of American bison: "The moving multitude ... darkened the whole plains." (Thus, the total number of large ruminants likely exceeded the 40 million mature dairy cows and breeding beef cows in the United States today.) When grazing animals are extirpated or otherwise disappear from a terrain, the ecosystem and all life that depends on it is affected. "When grazers are removed, grasses lose their competitive advantage and forbs and shrubs quickly become established," explains Environmental Resource Management professor J. P. Curry. Likewise, a Kansas State University study concluded "grazing by large herbivores is fundamental for functioning of tallgrass prairie ecosystems." In recent years, the recognition has emerged, including among environmental groups like the Nature Conservancy, that domestic grazing animals can be the indispensable surrogates for the vast herds of wild grazing animals that once roamed our continent.

Maintaining pastures and grasslands benefits the environment. Each blade and root of grass are threads in a dense vegetative carpet that protects the earth by guarding moisture and soils. On average, 90 percent of grass' bulk is actually below ground, in the form of long, filamentous roots—a tangled root network that stops soil erosion. "Grass prevents erosion by binding the soil," explain geologists John Rogers and Geoffrey Feiss in their book, *People and the Earth*. Adding, "replacement of grasslands with cropland in the American Midwest, its 'breadbasket,' has caused nearly one-third of the topsoil to be stripped by erosion in the past 100 years." Even wild areas of brush can fail to provide the earth a covering as protective as grass.

Cattle can play a vital role in maintaining or even restoring pasture and natural grasslands. The munching mouths of grazing animals aid grass areas in much the same way that mowing one's lawn keeps it lush and full. And their manure effectively recycles nutrients necessary for plant growth. Large grazing animals also disperse seeds of grass and other plants, which are spread efficiently when carried in the guts of cattle then deposited in dung and trampled into the soil. "Animals can be an effective and economical pasture renovator," says a University of Georgia publication.

Protecting grasslands and pastures has additional environmental benefits. A study conducted by the nonprofit Land Stewardship Project concluded that "permanent pasture for grazing livestock can be an ideal choice for minimizing [water] pollution." Pastures are also much better than cropland from a global warming perspective. Like forests (although to a lesser degree), grasslands act as carbon sinks, holding nine times the carbon of even well-managed cropland. "Pasture sequesters and holds carbon in the soil, in contrast with tillage-based systems, in which carbon is released each year." Pastures and grasslands also generate far fewer carbon emissions than croplands because they require little to no use of fossil fuel-based agricultural chemicals or machinery. And, a 2007 United Nation's report on livestock's effects notes, "[t]here is growing evidence that both cattle ranching and pastoralism can have positive impacts on biodiversity."

Finally, cattle grazing can be a wise use of resources. Exposés often claim that beef is invariably the most resource intensive food. The assertion is only partially correct because where and how cattle are raised makes all the difference. Cattle (particularly mother cows and breeding bulls) can and often do live entirely or primarily on naturally occurring vegetation. Their food is the most environmentally friendly agricultural input on earth: It requires no plowing, no planting, no watering, no fertilizers, no herbicides, no pesticides, and it doesn't even have to be harvested, dried, or transported. In other words, their food can be produced without any contribution to erosion, global warming, or pollution. Neither we humans, nor poultry, nor pigs can live off grass, rendering cattle invaluable intermediaries between natural, indigestible vegetation and human beings. We also cannot grow fruits, vegetables, or soy with so light an environmental impact.

Even in colder climates, well-managed grazing can provide nourishment to cattle for most of the year. Land and animals can be kept healthy by grazing animals on smaller sections of land and moving them regularly. An article in the *Dayton Daily News* described an Ohio cattle farmer's experience. The farmer planted winter-hardy grasses in his meadows and began to more actively manage his herd's grazing. He decreased the harvested feeds he was using and substantially expanded his grazing. "Grazing seven months a year and feeding hay the other five months was just way too expensive," the farmer explained. After shifting to year-round grazing, the farmer slashed his annual hay feeding from three tons per animal to less than one ton. "It saves us about $100 a cow by doing more extensive grazing," said the satisfied farmer.

And for the same reasons that cattle do well on grass, they can also survive on inedible (for us) by-products of food raising done for humans and other animals—(cellulosic materials like the stalks of corn, wheat, oats, barely, and rye). Their special gift of complex, multi-chambered digestive systems can effectively convert these by-products into sustenance. These unique cattle attributes give beef the potential—when herds are carefully managed and in appropriate geographies—to be the most environmentally responsible of all food derived from animals.

THE TRUTH ABOUT FEEDLOTS

With these points in mind, I'll turn to the three major areas of legitimate concern about modern beef production: inappropriate cattle feeding, drug abuse, and concentration of too many cattle in certain locations. Each problem is tied to feedlots (as opposed to cattle ranches). While working at Waterkeeper, every citizen complaint I received about beef production related exclusively to feedlots. Criticism rightly focuses there because feedlots hold cattle in crowded, unnatural settings. They generally provide the worst living conditions experienced by beef cattle, and are the segment of cattle raising most likely to cause pollution and public health hazards. However, as I would learn, not all feedlots are equal. A smaller, appropriately stocked and well-managed feedlot can have minimal environmental impact and provide cattle an acceptable quality of life.

A feedlot is just what it sounds like: an outdoor area where animals are kept in large dirt surfaced pens and provided harvested feeds. In the United States, most steers (and some heifers) leave the ranches on which they were born at a certain age and are trucked to feedlots. Feedlots then keep cattle for 150 to 270 days prior to slaughter. Despite it being outdoors, I call this an "unnatural" environment for cattle because they have no chance to graze and little opportunity for exercise.

As the beef sector's history suggests, feedlots were started for two main purposes. First, they made it possible to keep cattle herds on grasslands of varying quality. Cattle living on the range exercise constantly and eat a diet of natural forages. In other words, their lives resemble that of deer and elk. Correspondingly, their meat tends to echo wild game: lean and, (especially at certain times of year), having a strong, "gamey" taste that many people find unpalatable. As the Bible is replete with references to killing a "fatted calf," so, too, most Americans have long preferred their beef with some intramuscular fat ("marbling"), which is most easily achieved by feeding cattle grain in the months prior to slaughter. Feedlots make it possible to fatten those steers and heifers destined for slaughter while keeping the herds on the range.

In addition to mellowing meat flavor, feedlots smooth out the acute seasonal variations in supply. When cattle rely mostly on grass for their nutrition, they (again, like wild game) will be in peak condition only at the moment when grasses are the most energy-rich, i.e. just after the seedheads mature. Depending on location, this will be either a few weeks in spring or fall, making exclusively grass-based cattle raising highly seasonal and turning fresh beef into an ephemeral, Beaujolais Nouveau–like offering.

During the twentieth century, as Americans began processing, refrigerating, and transporting foods long distances and erecting hot houses for tomatoes, the idea of eating in tune with the seasons all but evaporated. Fresh meats, as fresh produce, were expected on a year-round basis. And of course, where there was consumer interest, there were always meat industry entrepreneurs ready to respond. Rising market demand along with eating preferences for fattened animals contributed greatly to the rise of the U.S. beef feedlot.

The advent of federal grain subsidies further tilted agriculture toward grain feeding and away from grass grazing. As we've seen, historically, keeping mother cows and still-growing calves and yearlings on grass had been cost-effective; hay and grain were reserved for over-wintering and for fattening mature animals bound for slaughter. But when grain subsidies entered the scene, in conjunction with the rising costs of owning or leasing real estate, grazing animals—*even on naturally occurring forages*—was no longer more economical than grain feeding.

A sea change in American cattle raising resulted. In the early twentieth century, young cattle were still being grown to full maturity—three or fours years old—on grass and *only then* put into feedlots, strictly for fattening before slaughter. However, by the 1930s, time spent growing on grass was often curtailed and many two-year-olds found themselves in feedlots. This trend accelerated over the decades that followed. By the close of the twentieth century, the idea of maturing beef cattle on grass had almost been abandoned. Those calves not selected for breeding herds were taken shortly after weaning and placed in feedlots (as young as six months old) for both growing *and* fattening. Cattle raising had morphed into a proposition with a singular focus on getting an animal to market as quickly and cheaply as possible. Said another way, when circumstances made it more profitable to *mature*, not just *fatten* animals on grain, the industry converted to doing so.

When Bill Niman first started raising cattle in the early 1970s, he raised them strictly on grass, supplemented with a little hay. Over time, he came to appreciate the value of judicious grain feeding for smoothing out the seasons and for creating marbled meat. Bill gradually altered his protocols to feed some grain prior to slaughter. Eventually, this developed into a small feedlot (originally located at the Bolinas ranch, then moved to another California ranch, then, finally to Idaho). The Niman feedlot was always used exclusively for fattening cattle from a network of ranches all following the same cattle husbandry protocols. At its largest point, it held around three thousand head of cattle. However, the Niman Ranch feedlot did not feed calves. According to traditional practices, which had been widely followed until the mid-twentieth century, the cattle were first matured (usually to around nine hundred pounds) on grass. Only then (as yearlings) were the Niman cattle put into the feedlot for fattening prior to slaughter.

I've visited the Niman feedlot in Idaho some half-dozen times, my first visit being a few months after leaving Waterkeeper. I had heard a lot of about the negatives of feedlots, so I approached it with a fair amount of skepticism. On that first visit and on every one following, I was struck by how good the animals looked. Their eyes were bright and clear; their coats were shiny. Since the feedlot pens were not crowded, they had plenty of room to move around. In the evenings, they would even lope about. Flies were minimal and the odor was almost imperceptible.

The Niman feedlot always made such a good impression because it is small and because it was extremely well run. Over the years when I visited the place, it was always managed by Rob and Michelle Stokes (no relation to contract hog grower Chuck Stokes), a couple who had worked for Bill for more than a decade. Both share Bill's deep commitment to humane animal husbandry and good environmental stewardship. The Niman feedlot never fed antibiotics or other drugs, never used hormones, and never fed animal by-products or anything other than hay, silage (fermented, chopped whole corn plants), grains, and trace minerals. It purchased much of its feed grains from nearby farms, and returned its manure to adjacent lands. I always prefer to see cattle on pasture, but from Rob and Michelle, I learned that a smaller scale, well-managed feedlot can be a hospitable place for animals and have little environmental impact.

However, in the mainstream beef industry, many feedlots became increasingly problematic around the mid-twentieth century. For one thing, they put younger and younger cattle—*calves*, really—into feedlots rather than allowing them to mature naturally on grass. Additionally, to increase per-animal profit by getting them to market faster, the industry rapturously embraced the latest products of biochemistry, especially growth hormones. Weight added by farm animals each day is referred to in the meat industry as "gain." After a 1954 study reported inexpensive, dramatic increases in feedlot gains, hormone use spread through the cattle industry like a California brushfire. Already by 1958, an estimated

70 percent of cattle in feedlots were being administered synthetic hormones. "No compound had as much impact on the beef cattle industry with as little information concerning its mode of action and effects in the animal body," an animal husbandry professor noted in 1958. Today, hormones for beef animals are even more popular, being used on an estimated 90 percent of cattle in feedlots, usually in the form of a time-release pellet, implanted subcutaneously in the animals' ears.

The European Union banned such hormones in 1989 over concerns about both residues in meat and hormones entering the natural environment from feedlot runoff and manure. Under ongoing trade pressure from the United States, the EU has repeatedly been forced to defend the prohibition. Thus, an EU scientific committee has twice undertaken thorough reviews of all available scientific studies. Both times, it determined that using growth hormones on farm animals poses unacceptable risks to the environment and human health, especially for young children. Likewise, the European Food Safety Authority advised the European Commission to retain its hormone ban. The body concluded that humans and wildlife are at real risk of exposure to hormones generated by cattle feedlots, stating that the substances had "the potential to alter the normal functioning of the endocrine system in wildlife and experimental animals." Among other studies, the committee cited U.S. research linking hormones from cattle feedlot runoff to endocrine disruption in fish.

Interestingly, because of human health concerns, the United States has banned the use of growth hormones in poultry and pork production (industries which, prior to the bans, also welcomed hormones with open arms). However, there are currently no signs of the United States disallowing hormones for beef cattle.

In the late-1950s, many cattle feedlots also began adding drugs to feed rations. Recall that around this time industrial poultry and pig operators started hurrying gain with pharmaceuticals. In cattle raising, the wave of drug use swelled in 1958 when the U.S. Food and Drug Administration approved the antibiotic aureomycin as an additive for cattle feed. Research documented that it and other antibiotics sped cattle gain. Around the same time, FDA also approved adding tranquilizers and other pharmaceuticals to cattle feed, all substances found to speed cattle growth.

The attitude of the era was unbridled enthusiasm for using man-made compounds in cattle husbandry. These developments were cheerfully noted by the (soon-to-be) Smithsonian curator of Agriculture and Mining, John T. Schlebecker who wrote in the early 1960s: "[T]he addition of antibiotics to feed promised to be one of the major scientific advances of all time. The practice promised unbelievable intensity of meat production." (Yeehaw!) Two pages later, he happily adds: "Scientists continued to try various combinations of antibiotics and hormones . . . and recorded impressive feed gains. Pellet implants, feed additives, pills, and shots, all had their day in one way or another . . . [A]ll of them did wonderful things." He concludes with a wink (and perhaps a nod): "At the rate of discovery and use in the fifties, biochemical pills and shots seemed about to replace food altogether."

The scientists of agribusiness just kept right on dreaming up substances that could be fed to feedlot cattle to lower production costs and thereby increase beef company profits. Adding urea, a nitrogen compound used as fertilizer, was a particular favorite. Urea is manufactured from carbon dioxide and ammonia, derived from coal or natural gas. Experiments completed by the late-1950s showed that urea boosted gain at little cost. However, they also revealed that it caused "varying degrees of kidney congestion" and was fatal to cattle when given in large doses. (One might also question whether feeding urea makes good use of energy resources.) But urea was effective at reducing feed costs, inexpensive and readily available, so feedlot operators sallied forth and began routinely adding it to cattle feed.

To the mix, add animal fats. We've seen that cattle are strictly herbivorous in nature. Yet a 1951 surplus of fats suddenly made them cheap and stimulated an upsurge of interest in adding them to cattle feeds. Research at several USDA experimental stations examined the use of "both edible and inedible animal fats," including a study at the Florida station that found that adding "5 percent raw beef tallow" to feedlot rations "increased both rate of gain and feed efficiency." These animal fats were, of course, the unmarketable by-products of slaughterhouses, which were thrilled to have a market for their waste. Slaughterhouses were also happy to unload their excess meats, organs, bones, and chicken feathers, all of which could be used as protein and mineral sources.

Because they were so cheap and abundant, animal by-products became common feed additives for both dairy cattle and beef cattle at feedlots (as well as poultry and pork) during the second half of the twentieth century. But in April 1985, a frightening sickness surfaced in the United Kingdom that would raise serious doubts about the wisdom of feeding meat to herbivores. The disease killed cattle after eating away at their brains until (when viewed under a microscope) they looked like sponges. It was named Bovine Spongiform Encephalopathy (BSE), and later dubbed "Mad Cow Disease." In 1988, scientists connected the illness to feeding cattle meat and bone meal coming from slaughterhouses and processing plants, which included parts from sheep infected with the disease scrapie. Before BSE was fully recognized, parts from infected cattle were being recycled back into the system and fed to other cattle. Mad Cow Disease swept across Europe, with two million cattle infected in Britain alone.

Within a few years, research revealed a disturbing link between the cattle disease and the debilitating and fatal human illness, variant Creutzfeldt-Jakob disease (vCJD). Not until 1997 did the United States finally outlaw feeding ground up ruminants (cattle and sheep) back to cattle. U.S. regulations still allow feeding cattle other meats (and allow feeding all meat and bone by-products to farmed fish, poultry, pigs, and pets).

And so it was that by the 1960s, the feed bunks of cattle feedlots had quite literally turned into America's dumping grounds. Yet another unsavory ingredient that began making its way into feed-lot rations around this time was poop of all stripes: cattle poop, chicken poop, and pig poop. Even though cattle on the range studiously avoid eating any kind of manure if possible, agribusiness researchers came to regard all kinds of poop as acceptable cattle feed (and poultry and pig feed, too). Experiments with manure feeding especially flourished in the 1960s and 1970s. "It can be concluded," summarizes a review of such studies, "that manure derived from cattle fed high-concentrate rations (dairy cattle, feedlot cattle) has good nutritive value and can be used as a forage substitute for . . . cattle."

Of course, farmers have historically prized animal manures for building the tilth (texture), fertility, and beneficial organisms of their soils. In appropriate quantities, manure greatly improves cropland and pastures. "The organic matter in manure is also valuable because it makes soil easier to manage, less likely to erode, and more likely to absorb water," notes a guide for smaller farmers published by the U.S. Department of Agriculture. "These are important benefits that inorganic fertilizer does not offer."

But industrial animal facilities, which contain animals by the tens and hundreds of thousands, will almost invariably generate excess manure for their vicinities. Because putting too much animal waste on land causes pollution, it is generally illegal (although, as Rick Dove's investigations have shown, such laws are frequently violated). The alternative to over-applying waste is manure hauling, which is often prohibitively expensive. The convergence of these realities created the intractable excess that triggered the drive to use manure as animal feed.

As in poultry and pork, the beef industry's obsession with lowering production costs has turned out to have many unintended (albeit foreseeable) negative consequences. For the cattle, many of the troubles come from putting young animals in crowded, stressful conditions. Just as babies and children are more susceptible to many human diseases, so calves are more vulnerable than mature cattle to various bovine illnesses. "Younger cattle generally have more health problems than older cattle," the Beef Cooperative Research Centre of Australia explains. "Younger animals have a less well-developed immune system and therefore are at greater risk."

Many feedlot health problems center around cattle guts. Mature cattle have complex, four-chambered digestive tracts. The rumen is the largest of the four compartments, containing billions of bacteria, protozoa, molds, and yeasts. These microorganisms, living symbiotically with bovines, perform most of the digestion, making it possible for adult cattle to convert into nourishment the cellulosic material that we mere single-stomached, nonruminants are not capable of processing.

However, the digestive tracts of calves are different. At birth, their rumens are small and their digestion actually functions like the single stomach (monogastric) systems of pigs and humans. By about three months of age, the calf's digestive tract begins to resemble its mother's. But traditional cattle husbandry wisdom teaches that only much more gradually is it fully converted to one that allows the animal to thrive on roughages and grains (a process called "hardening"), making it risky to feed young cattle a feedlot ration.

Digestive problems are common at feedlots, including acidosis and bloat. Acidosis happens frequently when cattle are fed the grain-rich diets of feedlots. Acids and glucose accumulate in the rumen, damaging the gut walls, decreasing blood pH, and causing dehydration. Bloat, a related concern, occurs when gasses normally produced and burped out during digestion instead accumulate in the first two digestive chambers (the rumen and reticulum). When cattle can't rid themselves of the gas, bloat occurs. Usually, this starts as a result of an acidosis episode. Because bloat is especially likely to occur when cattle are on rich feed, it's common at feedlots.

In the wacky dog-chasing-its-tail world of industrial agriculture, the prevalence of such digestive troubles at feedlots gives them yet another incentive to feed drugs. To control these problems, many feedlots add a category of antimicrobials called "ionophores," drugs like Rumensin (monensin) to daily feed.

Drugging daily rations is not the only way to control acidosis and bloat, it's just the cheapest. Bill has never raised cattle with drugs in the diet, including at the Niman feedlot, yet has always had very low rates of both acidosis and bloat. As one group of animal scientists noted, "Feeding higher amounts of dietary roughage, processing grains less thoroughly, and limiting the quantity of feed should reduce the incidence of acidosis, but these practices often depress performance and economic efficiency." In other words, as Bill oftens points out, drugs are cheaper and simpler than good animal husbandry.

An especially nasty disease on the rise in feedlots is Polio-encephalomalacia (PEM). Infected animals die suddenly after showing nervous symptoms that include ear twitching, loss of co-ordination, and blindness. They repeatedly fall down and roll over. Ultimately, they are unable to rise at all and go into convulsions. PEM, too, is associated with acidosis and with feeding high concentrate diets, especially to younger cattle.

Finally, there's the problem of coccidiosis, a widespread, sometimes fatal disease estimated to cost the cattle industry $100 million or more annually. Coccidiosis is caused by protozoan parasites and usually presents as acute, sometimes bloody, diarrhea. Research shows that coccidiosis is primarily a disease of young animals. Cornell University researchers note: "[T]he younger the animal, the more susceptible to infection they are." Coccidiosis "occurs commonly in overcrowded conditions" and "is transmitted from animal to animal by the fecal–oral route." For these reasons, feedlots holding younger cattle are uniquely susceptible to the spread of coccidiosis.

Not surprisingly, the pharmaceutical industry has rushed in with a panoply of products. The Merck veterinary manual, for example, advises avoiding coccidiosis (and also promoting growth) by adding synthetic anti-microbial agents (sulfa drugs) to feed. The guide recommends mixing drugs in dairy calves' milk-replacer beginning at two to four days of age, and adding drugs to calves' dry feed. These drugs are most effective, the guide notes, "when fed *continuously*."

Problems associated with feedlots tend to intensify as the scale gets bigger. Many smaller family-operated feedlots, holding fewer than 1,000 animals, are still scattered across the nation. But they are rapidly disappearing. Typical agribusiness feedlots these days, which are steadily replacing smaller facilities, hold 40,000 animals, and the biggest have almost 200,000. The most recent official census of agriculture showed about 100 feedlots, mostly in Kansas and Texas, having a capacity of 32,000 or more animals.

At that scale, it's difficult to truly monitor and care for animals, and impossible to know them as individuals. We've seen that chickens lost their individuality once they were housed and slaughtered by the tens of thousands and looked after by hourly laborers. Likewise, by necessity, giant feedlots reduce cattle to numbers.

"The eye of the master fattens his cattle." So opens the most enduring tome ever written on livestock rations, Henry and Morrison's *Feeds and Feeding*, a 1927 book still used and cited frequently today. The quote, and indeed the entire book, emphasizes the importance of attentive and skilled stockmanship in livestock tending. But at industrial feedlots the masters' eyes are focused on spreadsheets and are laid only briefly, if ever, on the animals. At a 90,000 head feedlot that Bill recently visited, hourly laborers were each assigned to keep track of 6,500 steers.

The enormous size of modern feedlots and the practices they employ raise the specter of serious environmental and public health concerns. The public comes in contact with feedlots through several possible vectors, most commonly via the meat they produce. Due to modern industrial methods, beef has the potential to be tainted with residues of multiple drugs, hormones, and antibiotic resistant pathogens. Researchers from the University of Maryland and the Food and Drug Administration randomly sampled 200 packages of ground meat in Washington, D.C.-area grocery stores. In 6 percent of the beef sampled, the scientists found *Salmonella* bacteria contamination. (Note that other meats were even worse: 35 percent of chicken, 24 percent of turkey and 16 percent of pork were found tainted with *Salmonella*.) Of the *Salmonella* strains isolated in the sampling, 84 percent were found to be resistant to at least one antibiotic.

A second vector for public contact with beef cattle feedlots is manure. And there's plenty of it: about 21,000 pounds per feedlot steer per year. Like industrial poultry and swine operations, most beef feedlots (about 83 percent) dispose of their waste through land application. The waste, which is treated as a solid, is scraped from pens and stored in huge piles. It is periodically transported to and spread on pastures or croplands. Small feedlots, especially those that eschew drugs and hormones, can dispose of manure in an environmentally appropriate manner. But, as with other animal facilities, when the operations become large and concentrated, pollution becomes unavoidable. "Underlying all of the environmental problems associated with [concentrated animal feeding operations] is the fact that too much manure accumulates in restricted areas," the U.S. Environmental Protection Agency (EPA) has stated. "Traditional means of using manure are not adequate to contend with the large volumes present at CAFOs."

EPA analysis shows that the vast majority of large feedlots—92 percent—have no land or insufficient land to safely apply the waste they produce. This means that nitrogen and phosphorous will likely end up contaminating streams, rivers, and groundwater. Feedlots with fewer than a thousand animals, however, fare much better, 84 percent having enough land for safe land application of their waste.

As with all concentrated animal facilities, everything put in cattle feed can potentially end up contaminating the environment. As noted by European Union scientists, hormones in feedlot runoff have been linked to health problems in wild fish populations. An ongoing study by researchers at the University of Nebraska expects to find that hormones used on feedlot cattle "will occur in cattle manure, will remain for extended periods of time in soil receiving cattle manure, and will under certain conditions be found in runoff from feedlots and fertilized soil." Likewise, an EPA report notes: "Although the hormone content of waste has not been systematically studied, a relatively large total mass of hormones is released yearly."

Antibiotics and other pharmaceuticals poured into feed bunks also wind up in feedlot waste, along with pathogens that have become resistant to the drugs. "Antibiotic residue may be found in animal by-products (manure and urine)," an EPA report states. "This waste may come in contact with humans, other animals, and surface and sub-surface waters through run-off and leaching. The concentrated use of antibiotics at CAFOs makes it more likely to have antibiotic residue and antibiotic resistant microbes in the vicinity."

Today's massive feedlots also cause serious odor and air pollution. Cattle feedlots emit ammonia, methane, carbon dioxide, nitrous oxides, and dust. Having been in and around a lot of cattle manure these past several years, I can say from personal and very close-up research that manure from cattle on pasture doesn't stink. It is, after all, just fermented grasses and clover. Truthfully, it has a sweet earthy smell that I must admit I find pleasing. Big cattle feedlots, however, are another matter. Like industrial poultry and pig operations, many are feeding a host of suspect ingredients and all are generating massive quantities of animal excreta that festers in a concentrated location. When nearing a large feedlot one cannot avoid noticing an eye and nostril stinging sensation. Many Californians are involuntarily familiar with a 100,000 head capacity feedlot close to Interstate 5. I've often heard people say, "You can smell it for miles."

And where there are mountains of poop, there are flies. Neighbors of large feedlots, people most affected by the odors, also have to contend with periodic plagues of pests. "There is evidence that fly populations remain a serious problem at many feedlot sites," acknowledges an Australian department of agriculture website.

Finally, rushing the growth of cattle and slaughtering ever-younger animals produces inferior meat. Hormones cause puffy, water-retaining animals, and their flesh is watery too. (This problem is exacerbated by the almost universal modern slaughterhouse habit of "overhead misting," in which moisture draining from hanging beef carcasses is replaced with water sprinkled from above.)

The flesh of younger animals also simply lacks flavor. Every farmwife worth her salt knows that older chickens make the best soup. So, too, the flesh of older cattle is recognized to have superior flavor. To compensate for the loss of flavor from killing younger animals, the beef industry devised a neat solution: use its PR muscle to promote *tenderness* as the most desired meat virtue. Conveniently, meat from young animals, while often bland, is generally quite tender. The idea that it's desirable to be able to cut your steak with a fork is as much a modern industrial invention as the manure lagoon.

To me, as a person now intimately connected with cattle raising, this whole array of unappetizing beef industry practices is incredibly frustrating. Each has a serious downside, and all are unnecessary. In more than three decades of ranching, Bill has never allowed any questionable practices to be used on any of his cattle or those in his ranching network. These are agribusiness shortcuts intended to do one thing: increase profits by lowering costs. Meanwhile, feeding slaughterhouse wastes and poop to herbivores and being addicted to drugs justifiably make the beef industry a tempting target for journalistic exposés. Such practices have lessened consumer confidence in all beef and given everyone involved in the beef industry a black eye—including cattle people who neither use nor condone such methods.

One last thing about criticisms of modern cattle raising: Animal activists have often attacked hot iron branding as cruel and unnecessary. I partially agree, but not totally. Branding animals on the face, (which slaughterhouses are encouraging lately to increase the value of hides), is certainly cruel, and I oppose it. I can't bear to look at cattle that are face branded. But branding does have legitimate and important purposes. Because cattle range over vast stretches of territory that is often many miles from the barn and abuts or overlaps land grazed by other herds, signs of identification are essential. Branding is a surefire way of putting a permanent identifying mark on an animal—one that cannot fall off or be intentionally removed. Just about every other ID mechanism falls short. Most alternatives (such as retinal scanning, or microchipping) are worthless for identifying cattle from a distance on the range, which is critical for running a ranch. That's why I consider branding defensible.

Nonetheless, I believe that most, probably all, of these aims can be accomplished by freeze branding, a technique that uses extreme cold, rather than extreme heat. In proper freeze branding, the hide is not damaged. Instead, super-chilled irons kill the pigment-producing cells in the hair follicles, causing the hair to turn white. (Some state laws currently require hot-iron branding and those laws need to be fixed.) On our ranch, we have converted to the freeze branding method, and it works fine. It appears to cause the animals no pain. Freeze branding does take more time to do and it costs more, but we think it's worth it.

A STRANGE HOG
FACTORY CONNECTION

In my regular travels to North Carolina, I noticed that confinement hog operations, such as Chuck Stokes', were increasingly keeping small herds of cattle. It seemed to have become trendy. At the Stokes' place, and elsewhere, the cattle were grazing fields at the very moment raw hog waste was being applied. This aroused my curiosity and concern.

By asking around, I found it was no accident many hog manure sprayfields are now dotted with cattle. While North Carolina was limiting (at least on paper) the quantity of hog manure applied to most land areas, the state was not regulating the amount of waste put on land used as "grazing pasture." In other words, by strategically placing cattle, like pawns on a landscape chessboard, hog operators could increase the amount of liquid manure they dumped on the land. Environmentally, this is absurd. Such a rule invites hog operators to put unsafe levels of manure onto grazing lands, increasing the chances of contaminating water and air.

From an animal stewardship standpoint, it's even worse. Grazing cattle on these tainted fields puts each of them at risk. High doses of nitrates are fatal to cattle. And, as with humans exposed to unsafe levels of nitrates, less acute exposures can lead to painful physical symptoms and spontaneous abortions in pregnant cows.

With some research, I uncovered several instances from the past few years where the cause of North Carolina cattle deaths was over-nitrified feed. An article from *The Charlotte Observer,* stated: "Toxic hay is killing cattle across the state at an unusually fast pace as farmers use up their winter feed supplies [of hay]." The article quotes a beef cattle specialist at a North Carolina diagnostic lab saying that his lab had seen about twenty whole herds die from eating toxic hay since the preceding fall. "The total number of deaths could be much higher," the article notes, "because not all livestock deaths are reported to state labs. Some of the hay tested that is being fed to North Carolina cattle found nitrate levels eight times higher than what scientists consider to be the safe limit."

Scientific research directly links this over-nitrified grass and hay to land application of hog manure. Forages with nitrate between 1 and 1.5 percent are considered "high risk" to animals while those over 1.5 percent are considered "severe or extreme risk." But in hog dense Sampson County, North Carolina, researchers found that such high levels were common. In Bermuda grass irrigated with swine lagoon liquid, "it was observed that under hay-cutting conditions the lower 3-inches of the forage contained 1.3 percent nitrate ion." While another study of hog lagoon effluent use found many cases of forages with high nitrate concentrations in hays harvested from hog confinement facilities. It noted that industrial hog operations now often spray swine lagoon effluent onto fields used to grow hay and found a connection between the practice and over-nitrified hay. "[I]n recent years, nitrate problems are becoming common in Bermuda grass and fescue forages due to large amounts of animal wastes ... being applied to pasture and hayland."

Disease-causing microbes are another concern. Columbia University has estimated that hog waste processed by current handling methods and sprayed onto land contains one hundred to ten thousand times as many pathogens as treated human waste. An EPA guide warns that hog waste may contain pathogens that "can cause disease in livestock, including *Coccidiosis, Cryptosporidium, Giardia, E. coli, Salmonella, Campylobacter,* and *Listeria.*" The list should also include bacteria in the *Brucella* family, which cause *Brucellosis* diseases in hogs, cattle, and humans.

From a livestock farming perspective, therefore, it's irresponsible to put cattle where you're spraying raw hog waste or to grow their feed where you're dumping excessive waste. I believe that no self-respecting cattle farmer would knowingly jeopardize his herd by grazing them on hog factory disposal grounds. Even if the animals survived, who'd want to eat the beef from those cattle?

However, if the cattle were merely shills for dumping excessive hog manure, one may be more willing to risk the deaths of a few cows or calves. Having a dead animal or an infected cattle herd might still be cheaper than hauling hog waste farther or building a waste treatment system.

To me, grazing beef cattle on hog facility sprayfields is yet another unsavory twist in the industrial animal factory system. It puts the only animals allowed a life outdoors on poisoned pastures.

LEARNING THE RANCHING ROPES

In my first few months on the ranch, I was unsure what my role here should be. On one hand, I felt some ambivalence about being part of a cattle raising operation. (Marrying a cattle rancher was one thing, becoming one myself was something else.) On the other hand, I was terribly interested in learning about this new place and what was going on around me. Fortunately, the ranch was not short on labor at the time, so there was no immediate pressure for me to get involved. Bill assumed I'd have no taste for ranch work and didn't ask much of me. Out of necessity, he occasionally requested that I check a gate or a water trough, and I happily contributed by performing these little chores. For the most part, though, I was merely an observer.

But still, I found myself longing to be in the company of the animals. To satisfy that desire and my curiosity about the ranch, I followed in my father's tradition and began exploring the ranch in long, daily walks. Part of my outing invariably involved watching and listening to the cattle. At first, I felt the need for vigilance and caution around these large, unfamiliar creatures, carefully keeping my movements smooth and deliberate to avoid the possibility of startling them and setting off a stampede. Sometimes I'd steer entirely clear of harm's way by just gazing at them over the fence.

The cattle were alert and full of life, with glossy black coats, bright eyes, and solid, shiny hoofs. Their appropriately proportioned bodies showed no trace of being manipulated by humans into distorted creatures, as has been done most severely to turkeys and chickens. Their legs were muscular and strong, their backs straight, their physiques firm and trim. Their udders were gently curving extensions of their underbellies, the dimensions of half a large cantaloupe. On dairy farms I'd watched cows struggling to walk as they dragged grossly oversized udders like pendulous, netted basketballs. In contrast, even our ten-year-old cows could (and often did) run and frolic among the grasses and wildflowers. When being led to a promising new pasture, they'd leap in exuberant celebration as though jumping over the moon.

My regular observations revealed our cows' quotidian patterns. Mornings, a cow rises early to graze while ambling forward in a steady, leisurely motion. The day's middle usually finds her relaxing in the shade of a tree alongside her herdmates, lying down and meditatively chewing her cud. After a few hours, she stands, stretches her limbs with deliberation, gets a drink of water, then continues her foraging trek into the twilight hours. All the while, she keeps one attentive eye on her calf, especially in its first few months of life. She will rush to the defense of her little one against any person or animal she considers threatening. I quickly learned that a mother protecting her calf can be the most dangerous animal on the ranch.

I especially enjoyed watching our cattle interact with one another and with other species—the horses, deer, dogs, egrets, and coyotes. The cattle were anything but the slothful, docile bovine "Elsie" characters known to me from advertisements and children's stories. At various times they were playful, aggressive, friendly, happy, demanding, timid, bold, and curious. Gradually, I learned how to interpret their body language and vocalizations.

The more familiar I became with cattle communications, verbal and non, the more confidence and comfort I gained around the herd. Soon I began conducting small experiments. One of my favorites was to simply sit myself down in the middle of the meadow they were grazing. No matter how often I repeated this, it invariably piqued their curiosity. Within minutes, a dozen or so cattle would encircle me with muzzles low to the ground, eyes wide, and brows quizzical. The boldest in the bunch would gingerly step forward and sneak a sniff of the oddly behaving creature seated in their midst. As soon as a nose's whiskers gently brushed me, the animal would shyly retreat several feet. After lingering and scrutinizing my purpose a few minutes, the clustered cattle would start to lose interest. Gradually, they'd drift away as the novelty of this inexplicable human behavior wore off. Through interactions such as these, I got to know the cattle and they got to know me. A mutual trust evolved.

These first several months on the ranch also became a time

of absorbing the language and wisdom of humans who work with cattle. I picked up pearls like, "always leave a gate as you found it," and "never get in between a cow and her calf."

Among my first lessons was that the word *cow* is not, I repeat, *not* a generic term. This one took a while to sink in. Like typical modern Americans, I'd spent my life using the word *cow* in reference to any bovine, regardless of age or gender. However, no farmer or rancher in America would use the word that way any sooner than she or he would use the word *hen* to refer to a rooster or baby chick. I had to learn that *cow* specifically and *only* refers to a mature female who's given birth to and weaned at least one calf. Ranchers use the term *cattle* to refer to a group and will always use a specific term (*cow, calf, bull, steer,* or *heifer*) when speaking of an individual. Such precision is critical for clarity of meaning.

What's more, this definition is not limited to the agricultural community. I found *cow* defined as a mature female by both my MacMillan and Webster's dictionaries. It seems that in an earlier era, few Americans would have failed to grasp the true meaning of *cow.* I think the loss of the word's proper usage is a sign of how disconnected we've become from the sources of our food. To build my ranch cred, I had to start straightening out my terminology.

Another invaluable lesson was how to carry myself around the cattle. Confidence is the key. These are big, powerful beasts. A fully grown bull easily exceeds two thousand pounds and a mother cow weighs well over a thousand. Either can seriously maim or kill a human if it really wants to. The saving grace is this: Like many animals that live naturally in groups, cattle have a clear sense of hierarchy and will rarely pick a fight with anyone they view as dominant. This hierarchy is referred to as the pecking order in chicken flocks while in cattle herds it is called the butting order. Every herd has a boss cow and several senior matriarchs, with each individual having its particular place.

We humans might view this kind of structure as rather undemocratic, but democracy and consensus are wildly out of place in the animal kingdom. Hierarchies maintain order and stability, minimizing conflict between group members, thereby reducing risk of injury or death. When the occasional fight does break out between cows or bulls, it can be a fearsome thing to behold.

The people raising cattle must always retain the clear alpha position in the herd. Dog whisperer Ceasar Millan instructs dog owners to firmly establish themselves as "leaders of the pack." The same concept is even more important in cattle ranching because humans are, evidently, much smaller and weaker than cattle. Failing to establish oneself as the head of the cattle hierarchy could be dangerous. Cattle are just as capable as dogs and horses at sensing a fearful or nervous person, and it arouses their suspicion and anxiety. A calm, self-assured person, however, inspires their confidence. It's not about instilling fear, it's about earning and maintaining the animals' respect. As an urbanite-suburbanite utterly unaccustomed to being around animals that literally weigh a ton, this meant that at first I had no choice but to feign self-assurance. Slowly, my confidence grew and turned real.

The time I spent in our herd of eighty-four mother cows also taught me that every animal was as unique as each child in a classroom. That may sound obvious but when an untrained eye takes in a sea of black cattle of similar size and background it struggles to distinguish one from another. However, as I came to know the animals, their physical differences became increasingly obvious. (I've had a similar experience with human identical twins.) Eventually I could recognize each of our animals with ease, even from a distance of thirty or forty yards.

Of equal note were their distinctive, individual characters. Some were trusting, calm, and self-confident, rarely flustered or afraid, whereas a few were skittish, wary, and shy. Most were a little of each. Some of the cows loathed canines while others were almost indifferent toward them as long as their calves weren't bothered.

One cow stood out the very first time I walked through our herd. She had a beautiful black and white brockle face, wide-set amber eyes, a calm, steady demeanor, and a long graceful tail that descended to a fluffy white switch. I'd particularly seek her out every time I meandered through the herd. She seemed quite unfazed by the attention, as she was about everything else, simply returning my admiring stares by looking me squarely in the eyes and rhythmically munching her cud like chewing gum. "I've found my favorite cow," I declared to Bill, and would regularly regale him with tales of her daily adventures. At some point we began referring to her as my "Girlfriend."

About four months after my arrival, I realized I wanted to be more than an observer here: I wanted to work on this ranch. By coincidence, I was already licensed to practice law in California and had initially assumed I would seek work as a lawyer for an environmental organization. But now I was intrigued by continuing my research and writing about the meat industry while simultaneously gaining a firsthand, real-world perspective of ranching. From months of informal study, I knew I was comfortable with the lives our animals were leading here and the way our land was managed. In fact, I felt proud of the lives our cattle experienced and the care they were provided, which contrasted so sharply with the lives of animals in the industrial system. I also hated being a helpless female in unfamiliar territory. Bill often traveled for business and I wanted to be able to cover the ranch in his absences. So, one day I made an announcement to Bill at our dinner table. "I've decided I'd like to learn everything it takes to run this ranch." "Really?" "Yup, and I want to start right away."

Bill was surprised by my pronouncement, but decidedly pleased. And he took me at my word. The very next day he encouraged me to come along on ranch rounds. In the months that followed, I helped him fix wood rail fences, patch broken water lines, and move cattle from one pasture to the next. Simultaneously, I became our ranch manager's shadow and lackey, doing any task she needed done, whether it was loading the truck with hay, searching for a wayward calf, or digging a fence post hole.

About six months into my training, I unofficially assumed the role of primary caretaker for our cattle and land. Our ranch manager lived twenty miles up the road on another ranch we leased, and Bill's company work took him away hours every day and sometimes for days at a time. I, on the other hand, was here every day. So it made sense for me to be the one making daily ranch rounds and dealing with routine matters. I liked my new job enormously, especially being outside for long stretches and looking after the animals. Even so, I felt a bit unsteady with such weighty responsibility, still feeling green in the art of ranching.

Initially, I tended to call our ranch manager or Bill several times a day. "I just noticed one with a white spot in her eye," I'd anxiously report from my cell phone. "Is the eye swollen or running?" "No." "Good. Which cow is it, anyway?" "Um, it's 301." "Oh, not to worry, she's had that for years." (How come I'd never noticed it before?) Clearly, I still had much to learn.

My favorite part of the job was just this sort of training in stockmanship. "It's best to see every animal every day," Bill would instruct me as we strolled through the cattle. "And when you're out here, don't just walk through them. Really look at each one. Check to see that their eyes are clear. Make sure none are limping. See how this one's lying down? We need to get her to stand up and walk a few paces. An animal lying down may be sick or injured, especially if the others are all standing and grazing."

I knew it would take years to fully digest the thousands of lessons of good cattle husbandry. I intentionally sought to master the knowledge by posing a steady stream of questions at our own ranch as well as at the ranches we visited. "I've never seen anyone ask so many questions," Bill often noted with an amused smile. Equally important to my education was my reading. The field instruction of my apprenticeship was supplemented with trade journals, historical books, and cattle husbandry manuals.

One of my books explained what to do in the unfortunate event that a newborn calf died. Skin the carcass, then place the coat on an orphan calf, it instructed. If done soon after the calf's death, the mother cow will usually adopt the skin-bearing orphan. It's called "grafting a calf." "Can you believe this?" I asked Bill with amazement the evening I read about it as I sat curled up by the fire with my cattle husbandry guide. "Oh, yeah. We've had to do that here several times." "God. I don't think I could do that." "You probably could." Little did I know that just three weeks later I'd be tested.

It was a cool, overcast morning, early. I was checking the pregnant cows and noticed that one had just given birth. Standing a respectful distance, I watched closely to make sure everything was in order. But I soon suspected it was not. The cow was standing in tall grass and vigorously licking her calf for several minutes. Normally by this point I'd see the calf shake its head and attempt to stand. Her calf lay too flat on the ground. Cautiously, I crept toward them. The mother glanced over at me and to my relief, did not turn aggressive. She actually seemed glad to see me, perhaps thinking I could help. Unfortunately, there was nothing I could do to save this calf, who had been born dead.

But I could try to provide her a replacement. At that moment, we happened to have a three-week-old male calf in the corrals. His mother had fallen ill and was not producing nearly enough milk to keep him healthy. For the past week, I'd been supplementing his mother's milk by bottle-feeding him several times a day. Within an hour of discovering the dead calf, and with its mother trailing behind us, the ranch manager and I carried its body to the corrals. Out of the mother's sight, we then removed its skin. Oddly, the task we were performing didn't feel at all grim or gory. Instead, I almost had the sense we were surgeons performing an emergency organ transplant operation. The calf's death was tragic. But if we carried out our jobs skillfully, the grafting procedure would give the grieving mother a new baby, the orphan calf a whole new lease on life, and make it possible for the sick cow to recover her health by relieving her body of the burden of nursing a calf.

Working quickly and quietly, we took the small black pelt and tied it to the second calf with short strands of twine. Then we carried him to the paddock that held the dead calf's mother.

We placed the calf on the ground and gently nudged him toward the cow. Then we stood back to watch and wait with our fingers crossed. For several minutes, the two animals danced uncertainly around the pen. Then the mother cow noticed this calf had a familiar scent. She craned her neck toward him and started sniffing him in earnest, from the rear end to the front. Bingo! This was now her calf. Seconds later, she began uttering soft cooing sounds that mother cows only make when talking to their own newborns. We were elated. By the end of that day, cow and calf were solidly bonded. She raised him with as much love and care as if he were her very own.

Some people say that in the grafting process the mother cow is being tricked into believing this is her calf. I suspect it's slightly different. I think the mother cow becomes willing to accept the calf, regardless of whether or not it's hers. Good mother cows have an intense desire to care for their offspring, especially in the hours and days immediately following parturition. The replacement calf fills a dark void left by the original calf's death. The way I view it, the mother cow is a willing participant in the substitution—she gladly becomes an adoptive parent. I feel a warm satisfaction whenever I see one of our grafted calves nurse from its adoptive mom.

My favorite cow, Girlfriend, has turned out to be an interesting story. She was raising a strapping steer with a black body and panda bear face when she and I first became acquainted. That summer our vet came out for the annual pregnancy checks. Girlfriend was the sole cow in the whole herd—*the only one*—that was not with calf. I was sorely disappointed. She was already eight years old and it was not unusual that her fertility would be declining. That's the sad moment when an old cow gets sent "to town," as it's euphemistically called on the ranch. Girlfriend, I felt, should be spared this fate and I decided to use my lawyering skills and plead for clemency. After a protracted, several week period of begging and negotiating, Bill agreed to let her stay another year. "Excellent! I just know she'll be pregnant next year," I blithely promised Bill.

Girlfriend had no calf to raise that following year, but she did earn her keep. Her exceptionally agreeable, serene temperament made her the perfect babysitter and I often observed her watching over a cluster of calves while their mothers contentedly grazed some distance off. We took advantage of her mothering nature and put her with the newly weaned calves for several weeks. She had a tremendously cooling, calming influence and she made a great mentor.

The second year that Girlfriend was declared "open" (i.e., not pregnant) by our vet, I had a harder case to make. "I'll buy her from the ranch with my own money," I offered when Bill was refusing to budge on the clemency question. "That's ridiculous." Finally, I wore him down and Girlfriend was granted another reprieve.

However, the next season, the day our vet arrived for pregnancy exams I knew I was facing an uphill battle. Girlfriend was now ten years old. It was an argument I probably could no longer win. As she marched forward in the line I turned to Bill and said, "I know. I've accepted that we can't keep her here forever." I tensely gripped the clipboard for recording each cow's exam results and awaited the inevitable verdict. "She's open, I already know," I sighed to the vet as he examined her. "Wrong. This one's got a big calf in her." Hallelujah! It truly seemed like a miracle. "You're kidding! Did you hear that Bill!?" He rushed over and gave me a big hug. We celebrated that night with a bottle of California sparkling wine.

Four months later, Girlfriend presented us with a healthy, beautiful red calf with a white patterned face. He was the biggest and sweetest calf of the bunch and Girlfriend, predictably, was the best mom. We dubbed him Isaac. The next year, she gave birth to an equally gorgeous daughter that we call Eve. Bill has now agreed that Girlfriend has won a permanent clemency and will be allowed to live out her days here on the ranch.

Over the years, Bill has taught me countless valuable ranching lessons, one being the importance of weather. In my past life, weather only mattered for figuring out what to wear. Not here on the ranch. When Bill rises, usually around 5 a.m., he starts his day by checking the internet or radio weather reports. It's the first thing he asks about when he phones a fellow rancher, whatever the purpose of the call. We carefully track every inch of rain with our rain gauge and record keeping system. We know when the sun will rise and set and the expected and actual temperatures, highs and lows. We even pay some attention to the moon, although not as much as some farmers and ranchers. It's something we'd like to learn more about in the future.

Weather affects what we do that day, that week, and what we're on the lookout for. Cattle eat less when it's windy or very hot and are more susceptible to foot disease and leg injuries when it's damp. High winds also mean downed tree limbs, which can take out fences. When the temperature swings radically between lows and highs, calves, especially, are vulnerable to respiratory diseases. Over the longer term, weather will also determine things like how much hay we store and how many females we keep. Ranchers are always looking for the perfect balance between buying more hay than needed and failing to have enough hay to make it through a dry, lean year. Every major decision on a cattle ranch will be based, in part, on weather.

On our ranch, hay makes up less than one percent of the animals' diet, but it's important. By late summer, the grasses here are dry and brown. That's when they hold the least nutrition. Even at that time of year, our cattle live primarily on grass, but we boost their diets with a little alfalfa hay. During those lean times, hay acts like a mineral and protein shake that supplements the animals' nutrition.

Other than that bit of hay, our cattle live entirely off this land. They simply eat the vegetation that occurs here naturally. We do no plowing, planting, irrigating, or harvesting. We use no fertilizers, herbicides, or pesticides. Our only manipulation of the land is the way we manage the grazing of our cattle and some limited, targeted mowing. The cattle drink water that comes from our reservoir, which is a catch basin for rainwater we collect and store year-round.

Almost nothing we do here depends on automation or machinery. When we (and 600,000 other Californians) lost power for five days two winters back it barely mattered on the ranch. Our cattle just kept right on grazing. There were no lighting or ventilation systems that shut down, no sewers that overflowed or backed-up, and, of course, no manure lagoons that burst. We have none of those. In fact, we use barely any generated energy to run the ranch. The only power we need is for pumping water to the troughs. Even that doesn't matter in the winter, though, because the cattle can find abundant water in puddles. So that power outage came and went and I'm sure not a single one of our animals noticed.

That really struck me because over the years I've collected a folder full of articles about whole herds of confinement animals dying in power outages. Unfortunately, it's a regular occurrence. As I first learned in Missouri from Scott Dye, the fumes are so noxious that a building full of pigs can suffocate within hours when such a facility loses power. But here, our animals don't depend on human generated power and aren't nearly as vulnerable to that or other human failings.

I love the ranching life and continue to learn new things daily by observation, experience, and study. Some days are longer than others, but they're all full and interesting. Once again, I've abandoned running, lap swimming, and biking. But this time it's not because I'm chained to a desk and computer. It's because I get more than my fair share of exercise working on the ranch. Quite often, especially during our calving season in the fall, the end of the day finds me bone tired, almost too exhausted to peel off my jeans and change into pajamas.

Living and working on our ranch has connected me with the natural world more than ever before. Every day I look for changes in the pastures, noting the growth or retreat of grasses, clovers, and vetch. I marvel at the wildlife that encircles us. Coveys of quail and single squirrels scurry across my path. Swallows swoop in figure eights over groups of grazing cattle. Bobcats stealthily slip into brush in the distance. Packs of coyotes and herds of deer trot through the fields before and behind me. Lone ospreys, hawks, and kites soar above. Pairs of ravens balance on the edges of water troughs. Vultures perch on fence posts like rising phoenixes, stretching their wings to warm them in the sun. The list is unending. In many ways, being so immersed in nature takes me back to my childhood days of exploring the fields near my home for hours on end. The difference is that now nature's cycles and rhythms tangibly affect my life, every day. As much as ever, I am in awe of nature's infinite beauty, wisdom, and power.

There is, I have to admit, a bit of Old Western movie glamour to the life we lead here, but just a touch. The feeling is strongest when I'm riding horseback through a grassy green meadow behind a slow moving group of cattle, sun beating on my cheeks and forehead. At other times, like when I'm fixing a fence in the driving rain or slogging through clammy, knee-deep mud to close a gate, the glamour is entirely elusive. Either way, I can't imagine anything I'd rather be doing with my life.

CHAPTER EIGHT

The Un-sacred (Milk) Cow

WHY THE TRUTH ABOUT
DAIRIES MADE ME BLUSH

Once I'd become a ranch hand and Bill's wife, lots of people calculated that the logical next step would be the return of meat to my diet. "Are you eating meat yet?" I kept getting asked. The question always made me smile. As a veteran vegetarian, the thought wasn't even in my mind. Not that I considered eating animal flesh morally wrong—I never had. Having spent much of my life studying and experiencing the natural world (and watching countless episodes of "Wild Kingdom" as a child and "Nature" as an adult), it has always seemed normal and natural to me that animals regard other animals as potential food.

During college, I did some wrestling with the issue. As a biology major, much of my time was dedicated to studying organisms and ecosystems. My studies reinforced my innate impulses that carnivorous eating was an integral part of natural cycles. Meanwhile, my freshman year roommate was dating a stern vegetarian, then morphed into one herself. They adamantly insisted that meat was both "disgusting" and "morally indefensible," barraging me (and everyone in their midst) with anti-meat propaganda. Beef was the worst, they claimed, calling it "the main cause of rainforest deforestation." While I didn't like the idea of forests being razed for my food, I've never been grossed out by flesh or blood (a trait that must run in the family since two of my siblings are now physicians) and didn't buy their argument that eating meat was unethical. So I resisted their peer pressure quite easily.

What their persistent nagging did accomplish, however, was to sow a seed. It got me seriously reflecting about what I was eating. Of course, I would not learn the intimate details of how animal-based foods were produced until years later. But that summer vacation after my freshman year I did mull over the question of what types of foods I should choose to sustain me.

After a few months of independently considering it, I decided that meat eating wasn't for me. Taking the life of an animal was a weighty responsibility, I concluded. While I'd been eating meat more or less out of habit, I'd never especially savored or craved it. At Vati's particular request, steak had been a Saturday evening ritual in our home throughout my childhood. I realized that for me, the steak held less appeal than a piece of rye bread soaked in the pan juices. And clearly, I could find equally nourishing alternatives.

One evening near the summer's end, I sat at the dinner table with my parents partaking in a meal that included fish. Midway through, I paused and put down my fork. "I've decided to quit eating meat," I declared with the gravity and finality of the president announcing a veto. (My parents, quite predictably, immediately expressed their worries that I couldn't stay healthy on a vegetarian diet.) That dinner was indeed the last time I've consumed any form of animal flesh. But I had no hesitation about continuing to eat eggs and dairy products. They were foods I relished and, in my mind at least, consuming them didn't conflict with my new resolution.

Thanks to my own off-putting introduction to vegetarianism, I've never pressured those around me about what they're eating. Nevertheless, I have to admit now that somewhere in the recesses of my mind I constructed a sly self-righteousness about my new diet. Although I did not condemn meat eating as immoral, I subconsciously felt a notch morally superior for not doing it. That is, until I began learning about the production methods of the eggs and dairy products I was buying.

This loss of innocence started at Waterkeeper, where I spent a fair amount of my time assisting activists battle against battery cage egg facilities and confinement dairy operations. I often provided them legal advice and relevant research. The concerns were unmistakably similar to those about industrial hog facilities: air and water pollution, pernicious odor, mistreated animals. It was increasingly clear that the eggs I was purchasing at the grocery came from hens cramped in wire cages who never saw the light of day and whose manure was probably liquefied and stored in giant, fetid cesspools.

Likewise, I was discovering that many of the cows providing milk for my butter, cream, and cheese had lives not unlike the hapless pigs contained in industrial facilities. A confinement dairy cow, too, would be forever deprived of feeling sun on her back, a fresh breeze on her face, or soft soil under her hooves. She would exist in a crowded metal building with concrete floors, liquefied manure lagoons festering a few yards off.

The idea of keeping any cattle continually in buildings especially bothered me. Unlike omnivorous chickens and pigs, all cattle are naturally designed to live entirely from slowly and methodically foraging vegetation. Bovines in the wild spend most of their waking hours in a state of ambulant grazing, walking an average of 2.5 miles a day, all the while taking 50 to 80 bites of forage per minute. Life in a confinement dairy promised a cow an environment and a diet that violated her very evolution. Faced with such facts, I was forced to admit to myself that for years I'd been ignorantly (and, indeed, blissfully) chowing on foods produced in places no better than the industrial hog facilities I was devoting my life to wiping off the planet.

I also realized for the first time that egg and dairy production are inextricably intertwined with meat. Laying hens end up in chicken soup (or, even worse, are simply thrown, still alive, into rendering processors). All dairy cattle become meat eventually (they once made up about one-half of U.S. beef), and virtually all veal comes from male dairy calves. "In fact," a dairy textbook points out matter-of-factly, in the United States "veal calves are merely a by-product of milk production." For some reason (one might fairly suggest willful ignorance), I had never given much thought to the fate of the animals creating my eggs and milk.

Now, however, these cold, hard facts were staring me in the face. They made it impossible to keep deluding myself that I could claim any moral high ground about my diet. I simply couldn't continue consuming eggs and dairy products and maintain any sense of moral superiority.

That's when I began to give it up. The moral superiority, that is. I still eat eggs and dairy, plenty. My favorite foods include cheese, butter, yogurt, and ice cream (with eggs not far behind). Where I decided to draw my line is where my food comes from, how it was produced. As I'll talk about more later, I have scoured the landscape in search of eggs and milk products from good farmers who treat their animals respectfully and use only traditional farming methods. I've also laid plans for raising our own hens and even a dairy cow or two. And, as with poultry and pork, I've witnessed the world of difference between traditional, smaller scale farming and industrialized agribusiness.

Undoubtedly, the most important element of my education about dairies has come from touring them. During the year and a half just after our wedding, I made a series of trips to Wisconsin dairies with Bill. He had been contacted by a group of progressive farmers there who had a lot in common with the farmers of Niman Ranch. They ran family owned farms that eschewed growth hormones and used only natural, herbivorous feedstuffs. The farmers were practicing a modern form of grazing (often called "rotational grazing"), which keeps animals in smaller fields (paddocks) and regularly moves the animals from pasture to pasture. Even in northern areas, where grazing provides only minimal nutrition in winter, these cattle are given access to the outdoors, providing them a life with conditions approximating that of wild grazing animals. These dairy cows do, however, receive ample supplements in winter and early spring, usually in the form of hay, grain, and silage (fermented forages).

The Wisconsin dairy farmers contacted Bill to tap into his substantial experience and knowledge about marketing foods created by traditional farms. Specifically, in an experiment partially funded by the state of Wisconsin, these farmers were considering getting into the veal business.

When we first received the invitation, I had never stepped foot on a dairy farm and knew precious little about veal, something Niman Ranch didn't sell. But the mere mention of veal made me queasy. While ignorant of the specifics, for at least a decade I'd been vaguely aware that veal calves in the United States were kept in small wooden crates, often tethered, and that eating veal was considered tantamount to wearing a coat made from baby seal fur.

"Why would you want to have anything to do with veal?" I pressed Bill. "Well, I'm not sure if we do. But here's the thing: dairy cows have to have a calf every year to keep their milk flowing. [This is called "freshening."] About half of the calves are male. Right now, there are three possible fates for each of them and none are very good." "OK. I'm listening. Go on." "Some are sent to the slaughterhouse when they're one or two days old. That's called 'bob veal.'" "Ugh. That's awful." "True. But the other possibilities aren't much better. Some calves end up at veal operations, where they're housed in confinement buildings, a lot of times in crates. The rest will be started out like veal calves then go to feedlots to be raised as beef." "So, I guess those are the lucky ones." "Most definitely."

I would later learn the magnitude of the issue. In one recent year, about 4.5 million male calves were born on U.S. dairies. Of those, 42 percent were immediately sent to slaughter; over half went to confinement veal operations; the remainder went to feedlots. Even those dairy calves sent to feedlots (which, ironically, are often marketed as "natural" beef) will never graze a blade of grass. And regardless of which of the three channels a male calf gets directed down, he is taken from his mother's side almost immediately and departs from the farm on which he was born soon after.

Viewed in that light, what the Wisconsin group was proposing had some appeal. Within their network of grass-based dairy farmers (called "graziers") they were experimenting with keeping male calves on the farm rather than shipping them off. Several different ways of rearing them were being tried by individual farmers in the network. All of the methods involved keeping the calves in groups and there was no crating or tethering. Meticulous records were being kept to determine actual costs and benefits of each approach. I agreed with Bill that we should go to Wisconsin and see it all for ourselves.

DAIRY'S CHANGING SHAPE

From our own advance research and a series of emails and phone calls with the folks in Wisconsin, we gleaned a lot about dairies before our first trip. I'd known for a long time that for most of the twentieth century, Wisconsin was *the* dairy state. Growing up in nearby Michigan, it was a matter of common knowledge. (Everybody knew that Wisconsinites were also known as "cheeseheads.") What I didn't realize was just how much dairy farming was being done in the state. In 1924, for instance, Wisconsin's dairy output was almost double that of the second most productive state, New York. (At that time, California's output wasn't even one third of Wisconsin's.)

For the first half of the twentieth century, Wisconsin's dairy farming and cheese-making economy bustled. A sizable farming population, hovering around 200,000 farms, churned out more milk for cheese every year. These were almost exclusively small and medium sized family owned and operated farms with fewer than 50 milking cows.

However, as industrialization crept into agriculture in the mid-twentieth century, Wisconsin's farming began to change dramatically. Between 1959 and 1997, the state lost more than half of all farms and more than three-quarters of its dairy farms. Simultaneously, dairy herds were getting larger. In 1945, the average Wisconsin herd had just 15 cows. By 1975, the average had risen to 34, and by 2002 it reached 71. It wasn't so much that small dairies were adding cows. The real reason average size increased so much is that large, total confinement operations were coming into vogue and replacing smaller dairies, pushing up the average. Today, the state "has several herds of over 1,000 cows and is building many more of this size," according to the University of Wisconsin.

A closer look at herd numbers is even more revealing. Figures from 1982 show that 90 percent of Wisconsin's dairy farms still had fewer than 100 cows, more than half smaller than 50. Even in 1994, not a single dairy in the state had more than 200 cows. However, by 2006—just twelve years later—there were 950 dairies with more than 200 cows, a quarter of those having more than 500 cows.

Contrast that with California, which toppled Wisconsin from its dairy throne in 1994. Recall that California didn't have much dairy farming in the early twentieth century. Correspondingly, the state had little established culture of dairy husbandry. When the mega-dairy was invented, it just came to California, set up shop, and became the norm. Already in 1982, only 4 percent of California's dairy cows lived on farms with herds of 50 or smaller. By 1994, California had 1,800 dairies having more than 200 cows. And by 2006, large dairies totally dominated. The state had more than 1,000 dairies that each had more than 500 cows.

Of course, size was only one of the differences between the

emerging industrialized dairies and traditional dairy farms. Early in American history, a cow was only milked during the strong grass seasons. "In the colonial days in America the cow was seldom expected to produce milk in winter time. She calved in the spring, milked fairly well on grass [and] dried up in the fall," explains a dairy history. Accordingly, fresh milk was consumed mostly in the warmer seasons of the year. Like cured meats, butter and cheese were methods of preserving milk during the seasons of plenty for the cold months to come, when the supply of fresh milk was scarce.

By the twentieth century, both cow milking and dairy eating happened year-round. Just about every American farm had its own cow, providing fresh milk for the family and surplus for sale. In 1939, the treatise *Dairy Science* noted: "Dairy cattle are more widely distributed on farms than any plant or any other animal except chickens, for more than 70 percent of the farmers keep at least one dairy cow." Cows typically passed most of their time on pasture and were provided supplemental nutrition (often oats) when they were milked. In northern regions, as weather chilled and the crops were brought in, dairy cows were put into barns for much of each day and provided harvested feeds grown on the farm, including hay, grains, and silage.

However, dairy farming's close connection to pasture was ruptured mid-century. As happened with farmers raising poultry and pigs, "in the decades following World War II farmers found that the use of relatively cheap energy, fertilizers, and pesticides, and greatly improved mechanization could improve farm profits." For many farmers, this meant taking dairy cows entirely off grass. "These inputs, which allowed greatly increased production per cow, were substituted for pasture in the production process," notes a paper on dairy farming by sociologists at Pennsylvania State University. "This trend gave rise to the predominance of confinement dairy practices on farms throughout the United States."

Yet the ethic still persisted that farmers should provide pasture whenever possible and a pleasant environment at all times. The dairy cow "does best" when she's on the excellent pastures of spring and summer, the 1956 treatise *Dairy Cattle and Milk Production* pointed out. "The maximum production reached at this season is possibly due largely to the excellence of the food, but at the same time the animal enjoys a moderate temperature and clean, comfortable surroundings," it continued. "There is an abundance of fresh air and sunlight, and the cow has perfect freedom of movement. Every effort should be made to duplicate these conditions in the dairy barn." The book also urged farmers that "in summer, all ages [of dairy cattle] should have the freedom of the pasture for a part or all of the time." Even in winter, it said, on hospitable days, "cows should be turned outside for a few hours." At those times when cows had to be kept in the barn, grooming was added to the herdsman's duties. "Dairy cows when kept confined in the barn most of the time [in winter] should be groomed daily."

It was also widely understood and accepted among traditional farmers that when put in barns, cows needed plenty of fresh straw or other absorbent bedding. Straw kept the animals comfortable, clean, and warm. "No satisfactory plan has been devised to do away with the necessity of providing bedding," noted *Dairy Cattle and Milk Production.*

However, the California dairies that took over in the 1980s and 1990s largely abandoned such dairying traditions. California's big new operations (many known as "dry lot dairies") were providing cows no grazing at all. These operations keep cows confined year-round, either on concrete or with a limited access to a dirt pen, often with only noncompostable sand as bedding. Feed is delivered to the animals via an automated conveyor system. Most of their time is spent in 4-foot-wide stalls. All concrete surfaces are curbed and sloped toward huge lagoons of liquefied manure.

In recent years, the largest dairies built in California (and other areas of the arid West) are enormous. Often referred to as megadairies, many newly constructed western dairies have more than

10,000 cows (one in Oregon reportedly has 60,000) and multiple large waste lagoons to match.

The operations and their lagoons are so prevalent in certain areas they have become major sources of air pollution. In California's San Joaquin Valley, a region that contains 2.5 million dairy cows, dairies are now the largest source of volatile organic compound (VOC) air emissions.

GETTING A CLOSE LOOK

I found such details of dairy's recent history discouraging. Yet there was a spark of hope in hearing about the graziers and I looked forward to getting to know their farms. Our first trip to meet them unexpectedly became part of our honeymoon. Bill and I had been planning a driving tour of the Upper Midwest in which we would share highlights of our earlier lives. First, we'd spend a week on the shores of Lake Michigan, where my family always passed several weeks of summer in my youth. We'd then stop briefly in Illinois to see some of my relatives before heading to Minneapolis to visit Bill's old haunts. There I'd be introduced to the home in which he grew up, the high school where he often cut class, the synagogue where he was bar mitzvahed, and the corner grocery store run by his family. Our driving tour would wind down with a stay at the Willis Free Range Pig Farm in Iowa, something Bill and I always regarded as a special treat. After being invited to see three Wisconsin farms, each trying a different method of on-farm veal calf rearing, we amended our original itinerary to make a several day detour on the way from Michigan to Minnesota.

It was early September and Wisconsin was at its finest. Warm, long days gently gave way to balmy evenings. We marveled at the stunning beauty of the rolling emerald hills cut by streams and fences and sprinkled with peacefully grazing cattle. Well-tended barns painted red or white stood as regularly posted sentinels of the landscape. We were getting antsy to get out of the car, talk with farmers and walk these fertile lands.

The first farm we visited was testing a simple but radical idea: allowing male calves to remain on pasture. All the calves in the experiment were fed mother's milk for the first thirty days. About half were then placed in a field with only minimal forage (having already been grazed by the cows) and were being bucket fed a milk-replacer (formula). We looked at this group first. The calves had plenty of room to run and play. They were friendly and curious and were clearly doing just fine.

Then we made our way up a hill to the pasture with the rest of the calves in the experiment. From this perch there was an awesome view of the surrounding patchwork of tree lines, farm fields, and meadows. The calves here were running with the herd in dark green clover and grass up to our knees, and suckling their mothers, making them the luckiest little guys in the whole dairy industry. Their mothers provided nutrition, companionship, and instruction. The calves had the chance to graze, romp, and play in the tall grass. We stood a good half hour chatting with our farmer guide and smilingly watching the peppy calves. The mother cows seemed to enjoy this chance to raise their young, too.

Of course, for beef cattle, mothers raising their young is the norm. For dairies, however, leaving the calf with its mother is utterly radical. Bizarrely enough, a dairy cow's milk has long been considered too valuable for her own offspring. Separated from their mothers within hours of birth, neither male nor female dairy calves are ever nursed or mentored. This is why a herd of dairy cattle in a field almost always consists exclusively of young adult or mature females.

Immediately separating mother and calf is not just the result of industrialization. Already mid-century, dairy farmers were warned against allowing calves to nurse from cows or even drink whole milk. "The dairy calf is almost always reared by hand because, as a rule, the milk of the dairy cow is worth more than the calf,"

Dairy Cattle and Milk Production noted in the 1950s. It had become normal procedure to separate calves from their mothers very soon after birth, feed them whole milk for several weeks and then follow that by six months of feeding skim milk, supplemented by hay and grain meal. (These days, female calves are generally fed rationed amounts of milk or milk-replacer for about six weeks then moved to solid food.)

The authors of *Dairy Cattle and Milk Production* would surely have condemned as foolhardy the calf rearing experiment Bill and I observed at this Wisconsin farm. Raising male dairy calves, even from skim milk rather than mother's milk, was not a profitable enterprise, they unequivocally pronounced. "Approximately 10 pounds of milk are required for every pound of gain made by the veal calf, but seldom will the selling price of the calf by the pound equal the market value of 10 pounds of [even skim] milk. Under common conditions every pound of gain on a veal calf is made at a loss." The economics of veal raising were clearly tricky, and we weren't yet sure if we wanted to get involved with it. Nevertheless, Bill and I thoroughly enjoyed every minute of our visit to that dairy.

The second farm stop on our honeymoon detour was a place we eventually came to know well. This was the Paris family homestead in Belleville, Wisconsin, twenty miles southwest of Madison. Pulling up the long gravel driveway we were merrily greeted first by an oversized yellow dog then by a rugged, equally welcoming strawberry blonde man in Carhartts. Bert Paris warmly shook our hands and invited us in for refreshments which, naturally, included slices of Wisconsin cheddar.

Bert and his wife Trish are leaders in the grassroots grazing movement. Over cups of coffee at the kitchen table, they described to us how it was they had come to that point. Years back, they'd dairied conventionally, using a hybrid of traditional and more industrial methods that's now common in Wisconsin. Their animals had grazed only minimally, nourished mostly from harvested grains and hay, and spent a lot of time standing on cement. The

Parises had always loved dairy farming but had been feeling it wasn't providing the quality of life they desired. They decided to try dairy grazing because it just seemed a better way to live.

"Once we had kids, I started thinking I didn't want to be spending so much of my time working with heavy machinery," Bert explained. Modern planting equipment, manure scrapers and spreaders, tractors, and feed mixers can be noisy and dangerous, not conducive to working with children. "I wanted farming to be something we could do together as a family," Bert continued. "I've found that some of my best conversations with my kids have been when we're out working with the cows."

Since converting to grazing, the Parises were indeed running the farm with a lot less machinery. The cows harvest most of their own food by grazing in the pastures. And they take care of most of their own manure by depositing it on the fields, conveniently returning nutrients to the soil and enhancing its composition. Although grazing is sometimes believed to require more labor, Bert and Trish's experience has been the opposite. They have reduced the size of their herd, lowering daily labor requirements, and they milk their cows only ten months of the year, enabling them to take family vacations. This overall reduced workload makes it possible to run the farm without any outside labor. "I'd rather manage cows than people," Bert remarked with a chuckle.

"Well, let's go see some cows," Bert suggested after we'd chatted a while in their kitchen. Bill and I readily agreed. Just steps behind the Paris home was the milking barn and the cows' resting area, which dairy farmers call a "loafing barn."

The barn, approximately 50 by 120 feet, was the shape of a Quonset hut but made from a heavy polypropylene cover stretched over a curved aluminum frame, which is called a "hoop barn" in farming circles. The dirt floor was thickly covered with straw bedding.

A small group of colorful cows, including several types of crossbreeds, was gathered near the loafing area. "I see you don't keep a lot of Holsteins," Bill remarked. Holsteins are those ubiquitous big black and white cows one sees on talking cow commercials and refrigerator magnets. They are by far the most common dairy breed in the country, today accounting for almost 95 percent of the U.S. dairy herd. "Naw. The Holsteins don't do very well on grass," Bert explained. "Not very thrifty. They produce a lot of milk, but they also take a lot of feed to maintain. Plus we've had more health and calving problems with them."

After leaning on the wooden fence together a while, we decided to take a look at the veal calves then walk out to a far pasture to see the rest of the herd. The calves the Parises were raising for the veal experiment were housed in two groups, some fed real milk and others milk-replacer. Both sets had a generously bedded roomy area in a hoop barn; one group also had access to a yard. All of the calves were bright-eyed and vigorous. They eagerly greeted us as well as the family dog, who lovingly licked the calves' noses.

It was now late afternoon and the light was turning soft and golden. We headed out toward the herd down a wide dirt trail surrounded on either side by lush meadows. "We move the cows a lot—once or twice a day, depending on the time of year—so they're always on fresh ground, and it gives the vegetation a chance to rest and grow back," Bert narrated. "That's really the secret to what we're doing here." Larger, fenced fields were carved into more specific grazing areas with the help of solar-powered portable electric fencing. Every morning, and sometimes again later in the day, someone in the Paris family would move the fences while the cows were in being milked. The cows would then return to a new pasture. The system was beautiful in its simplicity and efficacy.

The grazing method the Parises use also turns out to be good for the environment. Much of this benefit comes from the way grass covered areas function compared to croplands, which have much higher rates of soil erosion and carbon losses. Grazing farms also use less generated power. "[P]asture-based livestock farms can perform significantly better with regard to soil erosion, water quality, and carbon sequestration than confinement systems," researchers at Michigan State University have concluded.

After walking a ways down the path on the Paris farm, I looked up toward the back of their land and took in a memorable scene. The cows had decided it was time to come in for their evening milking and were diagonally traversing a large rectangular field. Led by the boss cow, an undulating line of about eighty bovines steadily ambled toward us in the golden glow. All of us just stopped to admire the parade. Bert put his arm around his wife's shoulder. Standing alongside Bert and Trish as they surveyed their migrating herd, I sensed their deep pride. They exuded contentment.

After visiting one more farm, which was as wonderful as the first two, Bill and I continued on our previously planned honeymoon journey. Driving northwest on I-94 we enthusiastically discussed what we'd seen and heard at the Wisconsin grazing farms. The places were striking—lovely to look at and full of life—where people and animals were thriving. These farmers were going directly against the dogma of industrialized agribusiness and flourishing all the more for it. In fact, they were doing better financially with a total grazing system than they had when farming conventionally.

This experience is typical, I later discovered. A family dairy in Wisconsin with "average or better management has a good chance of improving financial performance" by converting to grazing, research from the University of Wisconsin has found. One reason is that graziers are able to utilize existing infrastructure (like barns and silos) that cannot be used by large confinement operations. This lowers a grazing farm's overhead. In fact, several studies have shown that grazing dairies like the ones we visited require less capital investment and maintenance in buildings and manure handling equipment than confinement facilities. "By moving to pasture-based dairying, producers reduce capital expenditures for equipment since the cows harvest the majority of the forage . . . [Grazing] reduces the amount of waste that must be contained and simultaneously reduces the size and scope of waste management systems," a University of Missouri analysis found.

We also learned directly from the farmers that as graziers they spend less money on agricultural chemicals, pharmaceuticals, and vet bills. And their animals have more longevity. "Our cows are a lot healthier now, and they live longer," Bert had said.

For starting farmers, these cost savings can all add up to the difference between being able or unable to get into farming. "With reduced costs, lower capital investments and viable net incomes, grazing dairies may present more accessible start-up opportunities for beginning farmers," a 2007 study by Michigan State University concluded.

Bert spends much of his spare time inspiring and helping some of those young farmers. He's a vocal spokesperson for grazing and active in two grazier mentoring programs. On our next visit to Wisconsin we toured the farm of one his disciples. Steve was a lean, rosy-cheeked thirty-something. He tended a herd of pretty Jerseys, ranging in color from fawn to mouse, all resembling large deer. The breed comes from the English Channel Islands and is famous for excellent butterfat. Notably smaller than Holsteins, Jerseys weigh from 700 to 1,000 pounds when fully mature.

Over pieces of rich chocolate cake, which he'd baked in honor of our visit, Steve told us why he chose grazing rather than conventional farming. "I was working as a feed salesman and had been wanting to get into dairy farming for years. But I could never figure out a way to make it pencil," he explained. Until he learned about grazing, that is. "Grass-based dairy farming just makes so much more sense financially," Steve continued. Like Bert, he found many other advantages to grazing, too, especially in quality of life. He was now infectiously enthusiastic about the vocation. We drove away from Steve's farm thoroughly refreshed by the cake and from being around a young person excited about starting a life in farming.

This was our second visit to Wisconsin and followed our first by about six months, putting it in late winter and several months into my apprenticeship on our California ranch. The Wisconsin landscape was now white, brown, and leafless.

And this time we were surrounded by an entourage. It had occurred to Bill and me that we could assist the graziers by introducing them to customers who would willingly pay a premium for what they were producing. Accompanying us on this trip were people who could get the graziers' milk products into the mouths of hundreds of thousands of people across the country.

Having sampled them many times, I can personally testify that milk and cheeses from cows living on grass are in a league of their own. They have an earthy sweetness and complexity that one never finds in ordinary grocery store dairy products. As with a lot of foods, once you've sampled the goods, the commodity version tastes bland and lifeless by comparison.

Research demonstrates that grass-based dairy products are also more nutritious. In 2006, after conducting a comprehensive review of scientific literature comparing confinement animal raising with grass-based farming, the Union of Concerned Scientists produced a report called *Greener Pastures*. In its review of all existing nutritional analyses, the study concluded that grazing dairy cows produce higher levels of the beneficial fatty acids ALA (alpha-linolenic acid) and CLA (conjugated linoleic acid), an effect that was particularly pronounced in cheese made from pasture-based milk. ALA is believed to reduce the risk of heart attacks and CLA is believed to have many positive effects on heart disease, cancer, and the immune system.

Yet in spite of these notable benefits, graziers, by necessity, are generally selling their milk into the commodity market. In other words, it's mixed with confinement milk and they receive no premium for it, very much as Paul Willis had been getting no special price for his pigs until he met Bill Niman. Of course that is a loss for the farmers. But it's also unfortunate for the growing number of consumers who prefer foods from traditional farms. The group we'd brought along had the power to go a long way toward bridging that gap.

One person in our assemblage had a very different reason for coming with us. The sister of Diane Halverson, Marlene Halverson, also worked for Animal Welfare Institute and was equally dedicated to improving the lives of farm animals. The Halversons grew up on a diversified family farm with 25 milking cows near a town 35 miles south of the Twin Cities, in one of the area's many families of Norwegian descent. "Our cows were strong, healthy," Marlene remembers about the farm, where she and her sister performed chores like helping with the twice-daily barn cleanings, and keeping the cows in fresh straw bedding. "We used to bring the cows in from pasture too, with the help of our pony and our border collie, Puddles."

Like its neighbor Wisconsin, Minnesota has a strong tradition of family dairy farms with small herds. The Halversons farm was typical. In 1982 (when Minnesota was the nation's fourth most productive dairy state), 91 percent of the state's dairy farms had fewer than 100 milking cows and more than half had herds of 50 or smaller.

After pursuing educations and careers, the sisters had both moved back to the family farm to take care of their aging father. Marlene's many years on a farm with livestock make her (and Diane) unique in the animal advocacy community. As an avid student of the subject, she has toured farms all over the United States and Europe, and she has a depth of knowledge I've encountered in no one else. Marlene's a walking farm animal encyclopedia. I contact her whenever I have a question about animal welfare, especially if it relates to pigs or dairy cows, her particular specialties. Marlene resembles her petite sister in stature but is somewhat quicker to offer her opinion, a trait Bill and I did not at all mind. We invited her on the farm tour to get an expert's view of the various methods being tried to raise the veal calves.

At our first two stops, Bert's farm and that of his mentee, Steve, Marlene liked what she saw. Marlene gave the Paris' set up for veal calves an A-, preferring that all the calves would have some pasture access but otherwise well pleased with their care. At Steve's farm, she marveled at the physical fitness of his cows. "My gosh—look at that cow heading up that hill!" she pointed out about a cow that seemed to effortlessly cruise up a long, steep incline. Marlene appreciated the stockmanship evident at the two farms. These people knew and took an interest in each of their animals, showing pride in their herds and compassion for individual creatures.

But after our stop at Steve's farm, our day took an unexpected turn. Over lunch, the Parises mentioned friends with a modern confinement facility. The owners felt they ran a good operation and might be open to having us visit. "You guys have any interest in checking it out?" Bert tossed out to the group. We all did. Bert busied himself with making the arrangements on his cell phone. I was particularly pleased, knowing full well that what we had been touring up to this point was not the norm. This would be my chance to better understand the rest of the dairy industry.

We arrived at the confinement operation about an hour later. Of the couple that owned it, only the wife was in town. She had kindly agreed to show us around and greeted us at the building's front entrance with tousled hair. She wore jeans, a flannel shirt, rubber boots up to her knees, and rubber gloves up to her elbows. "I'm a bit of a mess," she started apologetically. "I just got done pulling a dead calf." Now that she mentioned it, I noticed her gloves and shirt were splattered with blood. "Well, come on in."

We all followed her into a front room the size of a large walk-in closet. Covering the walls were rows of shelves loaded with pharmaceutical boxes, bottles, and syringes. From there, we entered an employee area where our host paused to point to clipboards tacked on the walls and describe the record keeping system. Then we walked into an ultra-modern milking parlor. Here we halted for several minutes while our guide described the multiple benefits of the facility's equipment, which was largely automated.

Finally, we reached the area where the animals were kept. As it had from the outside, the building felt like a warehouse—built with metal walls and cement floors. The cows, all leggy Holsteins with pendulous udders, were standing around masticating their cuds. Their long black and white heads turned toward us as we walked in.

Marlene and I instinctively dropped to the back of the group so we could freely comment on what we were looking at. "God, I hate seeing cows spending all their time on concrete," Marlene said to me in a low voice. "It's murder on their feet and legs." Marlene knew of what she spoke. She had authored a comprehensive report about farm animal welfare for the state of Minnesota in which she'd addressed dairy cattle hoof and leg problems in some depth.

Marlene's report documents that lameness is at epidemic levels in the confinement dairy industry. Of every 100 U.S. dairy cows, there are an estimated 35 to 56 cases of lameness annually. Studies show that concrete floors are among its primary causes. "The main walking surfaces for cows play a major role in affecting the incidence of lameness, with concrete surfaces seen as a particular problem," notes the report.

The breeding of modern American Holstein dairy cows is another factor in rampant lameness. Cows with oversized udders are forced to walk with their legs tilted. Over time, this strained position is crippling. Additionally, by breeding dairy animals ever larger (for increased milk output), the industry has created an animal uniquely *un*suited for living on concrete. "Heavy body weight increases the likelihood that prolonged standing on hard surfaces will contribute to lameness," Marlene's report explains.

As our group moved through the confinement dairy building, I noticed something else unsettling: the cows' tails had all been cut off to stubs. I asked our guide why. "Well, it's just easier to milk 'em without their tails," she explained, adding, "My husband didn't like the idea, so I did it while he was away fishing for the weekend." I felt a warm rush of affection for her husband.

I have to say that a cow without a tail is a sad sight. At our own ranch, where cattle have their tails intact—the older cows' tails just reaching the blades of grass—I have even found myself admiring the beauty and grace of the cow's tail as she swishes it around. And I have often observed just how useful tails are to cattle. At certain times of year, cattle tails are in constant motion, flicking away flies and other insects that gather on their backs. Other than predators, which most cattle don't have to worry about much, flies are the bane of a bovine's existence. And in warmer months, confinement dairies have notoriously dense fly populations, making them places where cows are especially in need of their tails.

Yet this dairy was following the trends of the trade. Although the USDA does not keep official records on the practice, Marlene informed me that (as in the confinement pig industry) tail docking had become commonplace, especially among confinement dairies. The reasons given are convenience in milking and disease prevention.

However, there is little proof that tail docking, which is done without anesthetic, reduces disease—and there's plenty of evidence that it makes a cow's life unpleasant. In a 2002 article published in the *Journal of Dairy Science*, University of Wisconsin researchers noted that when it comes to tail docking, "no positive benefits to the cows have been identified."

As our group looped through the complex, I caught a glimpse of the manure lagoon out back. Within minutes, I broke away from the guided tour to take a closer look. I stepped through the rear opening of the building and walked several feet down a cement ramp. Before me, looking and smelling like the hundreds of lagoons I'd seen in Missouri and North Carolina, lay a fetid murky pond.

When I turned around to re-enter the building I noticed a dirty pile of bloody rags on the ground just outside the back door. My eyes focused on the heap and my stomach tightened as I realized it was not rags at all. It was the crumpled body of a newborn calf. I supposed it was the one that had been pulled from the cow just before our arrival. I felt entirely ready to get out of this place and on to our next grazing farm.

Our final stop on this wintry tour was at a comparatively large dairy in east-central Wisconsin, not far from the blustery shores of Lake Michigan. Although it had more than 450 milking cows, the dairy was a grazing outfit like the smaller farms we'd been visiting. The farm was run collaboratively by several members of a fifth-generation German family, the Klessigs—brothers Karl and

Robert, sister Elise, and brother-in-law, Jerry Heimerl, with the benevolent oversight of their elderly father, Edward. Pulling into the farmyard, I was immediately struck by a stone and masonry silo that looked straight out of the Old World. Mounted on a nearby red barn was a neatly painted wooden sign that read: "Saxon Homestead Farm, est. 1850." A white, blue and green hex sign, surrounded by six horseshoes, was painted on the red barn door. The entire Klessig clan greeted us in the driveway.

Soon we embarked on a proper tour of the farm, which was humming like a beehive. A few paces left of the driveway was a sweet Collie dog watching over her litter of playful young pups. Down a narrow path to the right was a small stone and timber outbuilding that held an enormous caldron of maple syrup bubbling over a wood fire. Tapping sap from the farm's maples, followed by several days of boiling and bottling, was a yearly late winter tradition for the Klessig family.

To the left of the little building was a clean, sunlit hoop barn. On both sides of a center aisle were rows of young calves. Each was separated from neighboring peers by wire panels, in an individual pen thickly bedded with fresh straw and enough room to move about. These animals were not part of the veal experiment. The Klessigs, in their cautious German way, were still deliberating over whether to get involved, fearing that veal calf rearing could not be done without financial losses. Also, the Klessigs already had a pretty good solution for their male calves: they went to a neighbor's feedlot to be raised for beef.

I stayed close to the calves, stroking them for several minutes. These gentle creatures with silky hair and doe eyes soaked up the attention like puppies. I found it impossible to tear myself away. Meanwhile, our group drifted out toward a state-of-the-art milking parlor and an old wooden barn. Eventually, I followed.

By the time I caught up again, the group was in the barnyard. They were intently observing a cow a few yards away preparing to give birth. "Don't you need to, uh, get in there and assist her?"

someone asked, a bit nervously. "Nope. Our cows don't usually need any help from us," Karl Klessig calmly responded. And indeed, with all of us transfixed, the mother cow did just fine on her own. After pushing out the calf, she stood, licked off her newborn, and nuzzled it to stand and take its first meal from her udder. Everything was very much in order.

A bit later the Klessigs led us toward a sizable loafing barn. A woman in our group who knew little about pasture-based farming commented, "Aha! So that's where you keep the cows during the winter." She had inadvertently insulted our hosts. "Not at all!" blurted Robert Klessig with more than a trace of indignance. "Our cows are never confined. They can go in the barn whenever they want, but they almost always choose to be outside. Only a blizzard or cold driving rain keeps them indoors."

We walked on and moments later the herd appeared before us on the gently sloping snowy white landscape. The cows were of mixed origins and various colors. Among them were several Normande cattle, a French breed I'd never seen in the United States, whose milk famously makes the best Camembert. The cows quietly rummaged around in the snow and milled about. Some were lying down. Our hosts explained that the animals' diets were heavily supplemented during the winter but that they benefited from the daily exercise and fresh air. The herd seemed quite contentedly occupied.

As our tour wound down, the Klessigs welcomed us into their home, an early twentieth century American two-story four-square. We entered by the back door, were ushered into the dining room and invited to take seats around an ample wooden table. There Elise served us a feast that could have been unflinchingly presented to Queen Elizabeth. Cheese, I hardly need add, was prominently featured throughout. As we prepared to depart, we were thanking our hosts for their graciousness when they insisted we each take home a bottle of maple syrup. We left with contented smiles and full bellies.

Shortly after our visit, the Klessig family was recognized

for outstanding environmental stewardship. In the 1930s, father Edward had taken a course at the University of Wisconsin from world-renowned naturalist Aldo Leopold. The Klessig children say it had a profound and lasting impact on their dad. It helped instill an environmental ethic that he evidently passed on to them. Wisconsin's River Alliance recognized the Klessig family for doing "extensive water monitoring of the runoff from their fields" and for decades of commitment to land stewardship and water quality.

Bill and I stayed in touch with the Wisconsin graziers. We visited the Klessig and Paris farms again, about a year later. On that and every visit, they impressed us with their unfailing knowledge and professionalism. But we jointly reached the conclusion that it was not the right time to proceed with a veal venture. Bill and I had a concern, shared by several of the farmers in the veal rearing experiment, about feeding the calves anything but pure, unadulterated milk. Veal formula universally contains animal by-products, like beef tallow, lard, and blood meal, which formula manufacturers buy from slaughterhouses and rendering plants. As early as the mid-1950s, calves were being fed blood meal as part of the ration. Formula is much cheaper than real milk and is essential to making veal profitable under current market conditions. We saw no immediate way around the problem and decided to put the veal rearing idea on the shelf, at least for the time being.

Within two years of our final visit with the graziers, both the Paris and Klessig families had started cheese-making enterprises. For the Klessigs, this had been a long-time dream. Sadly, family patriarch Ed Klessig passed away about a year after our last visit with the family. He had been an indefatigable proponent of the family's cheese-making venture, according to his son-in-law Jerry. "He was an incredible man," Jerry said to me on the phone recently. "He fully supported the Saxon Homestead Creamery, every step of the way."

Although we didn't go ahead with the veal project, Bill and I have never had any regrets about the time we spent looking into it.

We enjoyed learning about dairy farming and, even more so, getting to personally know some Wisconsin graziers. Their movement is a rare yet intensely bright spot in modern agriculture. The farmers we met are leading the way. Modern grazing dairies represented just 7 percent of Wisconsin dairy farming in 1993 but jumped to 22 percent by 1999 and continue to spread. Unlike most other governmental bodies, the state of Wisconsin deserves some serious credit, too. It has actively supported the movement, recognizing, it seems, that grazing can distinguish Wisconsin dairy farmers, and that grass-based farming holds great promise for the future.

THE AMERICAN MILK FLOOD

For thousands of years, humans have guided the formation of distinct varieties of domesticated animals. Although there were several well-characterized breeds of domestic cattle by 2500 B.C., it is a recent idea to doggedly push farm animal physiology toward one sole purpose, such as milk production. In fact, until the late eighteenth century, the distinct types among cattle were mostly formed by natural conditions like climate and topography, and by the introduction of new cattle types due to human conquests and migrations. The idea of one cattle type for meat and another for dairy would surely have seemed impractical and probably nonsensical. That began to change when, around 1770, "a great interest was aroused in England in the improvement of the quality of cattle and other domestic animals of Great Britain."

Of course, breeds result from such human interests. Nature generates seemingly infinite variety. Some individuals in a species will be larger, some faster growing, some hairier, some more fertile, some more milk-producing, and some fleshier, with all sorts of combinations of these and manifold other traits. Humans, by selecting and breeding certain animals, form creatures with particular characteristics emphasized according to human needs and preferences.

With the advent of artificial insemination, which became the norm in the dairy industry starting in the mid-1950s, humans exerted even more precise control over farm animal breeding and dramatically narrowed the gene pools.

We've seen Americans in the late nineteenth century creating colorful chicken varieties, ultimately designating certain breeds for egg production and others for meat. The British impulse with bovines in the latter part of the eighteenth century was comparable. They began developing fleshy cattle for beef, and cattle with highly productive udders for milk.

Nonetheless, for most of American history, the idea persisted that animals should be bred for multiple purposes. Before the days of tractors (and still today in many parts of the developing world), cattle were essential beasts of burden for the farm, pulling carts and plows. Thus, it was desirable that cattle be fit for at least three major purposes: work, meat, *and* milk. An example of such a triple-purpose breed is the Devon, which was developed in Devonshire, England and probably first brought to the United States in 1817. In the early twentieth century, Devon cows weighed around 1,000 pounds and produced as much as 5,000 pounds of milk annually. Another example is the Brown Swiss, which had been developed over thousands of years as a hardy triple-purpose breed by the people of Switzerland.

However, early in the twentieth century, Americans showed strong signs of tossing aside multi-purpose breeding. For instance, although originally categorized in the United States as a dual purpose breed, "in 1906 the Brown Swiss breeders in America decided that their stock should be classed as a dairy breed, and indicated their intention to develop it with this purpose in view." Additionally, each of the major dairy breeds extant in twentieth century United States (Holstein, Jersey, Guernsey, and Ayrshire) originated from Europe and was emerging around this time with a distinctive "American type." In each case, the American version had been bred to a more "extreme dairy" body.

Yet not until the second half of the twentieth century, with dairy farming's industrialization, did single-purpose cattle fully dominate American agricultural thinking. In the 1950s, one could still find a U.S. farm book containing the following discussion about specialized cattle breeds, evidence that the matter was not yet fully settled: "Various classifications of cattle have been made, some based upon geographical distribution (as lowland and mountain cattle), others upon their anatomy (especially the shape of the skull), and others as to their domestic use. Today most American writers arrange them according to the economic value as beef breeds, dairy breeds, and dual-purpose cattle."

Breeding for the extremes and for a single purpose can have dire consequences for the animals. We've seen with chickens that males hatched to hens of egg-laying breeds became mere refuse. Turkeys bred to have swollen breasts lost their ability to breed and even to walk. Likewise, it was single-purpose breeding that reduced male dairy calves to unwanted by-products.

The extreme dairy body type involved unnaturally large dairy cattle and grossly inflated udders. In addition to lameness, large body size also means oversized calves, contributing to frequent problems for both mother cows and newborns in birthing.

Exaggerating the udder means that the body is devoting an unnatural portion of its energy to one body part to the detriment of the rest. That partly explains why single-purpose dairy cattle have a battery of health problems and, not surprisingly, became quite unappealing as meat animals. They still end up slaughtered for beef, but their economic value as meat animals is low. This makes it generally a losing enterprise to raise male dairy calves for beef (except when beef prices are unusually high). It also means that dairy farmers get little recompense for the cows (adult females) they send to slaughter. (I suspect this is part of the reason why some of the worst slaughterhouse abuses seem to happen at plants that specialize in the slaughter of old dairy cows.) Thus, the extreme dairy body has turned into a proposition with multiple serious downsides.

Corresponding with the rising emphasis on milk production at the dawn of the twentieth century, the social importance of milk and dairy products was soaring to new heights. As manufactured milk products were invented and expanded, their producers marketed them with a vengeance. Cheese was "promoted by strenuous merchandising campaigns;" ice cream was "aggressively merchandized" by its manufacturers; and powdered milk was promoted by a "comprehensive campaign" of the American Dry Milk Institute. (Years later, promotion became mandatory for dairy farmers when a fee—called a "check-off"—of 15 cents for every hundred pounds of milk produced was instituted. The dairy check-off promotions include the ubiquitous "Got Milk?" ads.)

Some advocates were even zealously touting milk production and consumption as essential to the realization of America's Manifest Destiny. "The keeping of dairy animals was the greatest factor in the history of the development of man from a state of barbarism," reportedly declared Dr. Elmer V. McCollum (who discovered vitamins A and D). He was also said to have pronounced: "Without the continued use of milk, not only for feeding of our children but in liberal amounts in cookery, we cannot as a nation maintain the position as a world power to which we have risen."

From such a perspective, increasing milk output wasn't about making more money: it was a humanitarian and patriotic duty. Reading about the drive for ever greater milk production calls to mind America's race to the moon.

Dairies across the nation focused a new vigor on just how much milk each cow generated per annum. Breeding and feeding were the two primary tools in pushing up her production. In 1850 each milked cow had fed just over three persons. In 1909 Americans were consuming an average of three pounds of cheese per year and figures from the following year show that a cow produced an average of 2,902 pounds of milk per year.

By 1950 each dairy cow was being asked to feed six people, more than double what had been expected a century earlier. Her milk production average had risen to 7,300 pounds (notably varying by breed, with the Holstein average highest at 8,750). U.S. cows collectively produced more than a billion pounds of cheese and annual per capita cheese eating had gone up to almost eight pounds. These increases were impressive. But they were paled by what was to follow.

By 2005, almost all U.S. dairy cows were Holsteins, the most quantitatively productive breed, weighing on average 1,400 pounds (up from 1,250 pounds earlier in the century). They gave an average of 19,951 pounds of milk a year. In other words, over the course of the twentieth century, the dairy cow's average production was increased by *a factor of seven.* Today, total annual cheese production exceeds 9 billion pounds and Americans are eating an average of 31 pounds of cheese a year, a number expected to continue rising.

"Improved genetics accounts for about half of the annual increase in U.S. per-cow milk production," states a 2004 University of Wisconsin publication. But one does have to wonder: What is meant by "improved" when dairy cows have trouble walking due to oversized udders and epidemic levels of lameness, and are routinely sent to slaughter before the age of four, several years before even reaching their peak milking age of six to eight?

Along with breeding, feeding has long been the other key factor in the equation for maximum milk production. With most animals off pasture, agribusiness scientists sought cheap feed sources and combinations for increasing a cow's milk output, much as they sought formulas for cheaply fattening steers. Here again the strictly herbivorous nature of the bovine was judged irrelevant.

Among common feed additives listed in the trade's official *1994 Feed Industry Red Book* were the following animal by-products: blood meal, hydrolyzed [poultry] feather meal, fish meal, meat and bone meal, and poultry by-product meal [necks, feet, undeveloped eggs, and intestines]. Each provides concentrated sources of protein. The high fraction of certain types of proteins is said to "make animal-marine protein by-products particularly attractive in diets for lactating dairy cows."

We've already seen how feeding meat and other animal by-products to cattle was connected to the spread of Mad Cow disease. Dairy cattle are at even greater risk for Mad Cow than beef cattle because they are given more total feed in their lifetimes than beef cattle. "Dairy cattle are also more likely to receive protein supplements in their feed (to support their milk production) than beef cattle," the Consumers Union has noted. For these reasons, it's not surprising that about 80 percent of the Mad Cow cases in the UK occurred in dairy cattle. As noted previously, in 1997, the United States restricted feeding ruminant by-products to cattle but still allows other animal by-products.

Such animal by-products have been liberally fed in the dairy industry. Typical feeding rates for blood meal, hydrolyzed feather meal, fish meal, meat and bone meal, and poultry by-product meal are between a half-pound and two pounds per cow per day, a University of Wisconsin article reports. However, the article adds that, as a practical matter, these levels may have to be reduced. At a certain point, dairy cattle simply reject the feed based on taste. In particular, blood meal, which is "coagulated packing house blood which has been dried and ground into a meal," is notoriously unpalatable to cattle. It seems that when the portion of animal by-product is high enough, the natural repugnance bovines feel toward eating animal flesh overwhelms their hunger and moves them to go on a collective hunger strike.

The dairy industry has also employed growth hormones to increase milk production. The most widely used hormone is called Recombinant bovine somatotropin (rBST), an injected compound manufactured by Monsanto. The most common users of rBST have been larger dairies. For example, University of Wisconsin figures from 1999 show that more than 70 percent of Wisconsin dairies having more than 200 cows were using rBST on their cows while its use was much less common among smaller dairies.

Due to consumer safety concerns here and abroad, the use of hormones on dairy cows has not become as widespread as once expected. Such concerns contributed to the European Union's decision to outlaw the use of growth hormones on dairy cows. What is often forgotten is growth hormones' effects on the animals. "[Their] use results in high milk yields and higher levels of mastitis, lameness, reproductive disorders, and other problems such as those at the injection site."

Although the skyrocketing milk production is almost universally celebrated by the dairy industry, it has come at tangible costs like these for the cows and those who take care of them. "The production of a liberal amount of milk is a severe tax upon a cow," *Dairy Cattle and Milk Production* advised already in 1956, a year when average milk production was just one-third what it is today. Signs were emerging that breeding for the extremes might cause problems for the animals. "A well-developed udder is of first importance in selecting any dairy cow," the book urges. Adding, however, "Defects are common in all breeds."

Udders, cows' four-quartered mammary glands, have ballooned into the industry's biggest health problem. Mastitis, an inflammation of the udder caused by pathogens entering the teat, is considered "the most economically imposing disease facing dairy producers in the United States," causing an estimated $2 billion in economic losses annually. "High levels of milk production are widely associated with high levels of mastitis." Scientific investigations have also shown a connection between rising milk yields and increased incidences of other maladies, including leg disorders and lameness.

Confined dairy cows' only exercise is walking short distances (on cement) between areas for lying down, eating, and being milked. This lack of activity adds to their health problems. Like people, cattle who regularly exercise are more agile and less likely to fall (a particular hazard on slippery, manure coated concrete floors). Animals on pasture move like athletes, says David Zartman, an animal sciences professor at Ohio State University. "They're more sure-footed." That's especially significant for milk cows, Zartman says, because they have so many foot and leg problems on concrete.

Perhaps the most tragic aspect of to-the-moon milk production is that it is wholly unnecessary. In fact, in many years heightened production has caused gargantuan dairy surpluses. "The capacity and willingness of the U.S. dairy industry to produce milk has equaled or exceeded its capacity to market milk and dairy products during nearly all of the past century," notes animal sciences professor Carl E. Coppock. And a Cornell University publication warned in 1983: "The [surplus] problem is not temporary but real." In the 1980s, dairy surpluses were of such grave concern that in 1986 the federal government even paid dairy farmers to slaughter almost 10 percent of the U.S. dairy herd.

Responding to the ongoing dairy excess, the federal government has long purchased cheese, butter, and nonfat dry milk under a "dairy price support program." The products are stored and, to the extent possible, funneled into domestic or foreign feeding programs, including school lunches. What can't be poured into some type of program is put into storage. At times the dried milk surplus has been extreme. "In 2002, storage costs alone for the powder peaked at $2.3 million a month," said Michael Yost, associate administrator of the USDA's Farm Service Agency. In 2005, the storage costs had dropped to $360,000 a month, comparatively, a bargain.

From an economic perspective, the industrialization of dairy farming, which has made it tremendously capital-intensive, deprives farmers of their nimbleness. As economists like to say, the system is now *inelastic.* The farmer can no longer respond to changing market conditions by increasing or decreasing herd size or milk output. Instead, (just as we've seen with poultry and pig contract growing), farmers that have opted into the industrialized system are now servants to massive debt. They have so much money invested in their buildings and equipment that they must continually generate maximum revenue just in hopes of making their monthly payments. As Marlene Halverson has put it, "the dairy cow must be worked to the limits of her productivity to meet the high costs of labor, housing, machinery, and other costs."

My journeys into the dairy industry have been tremendously eye opening about the food I eat. I was especially dismayed to learn that some "organic" dairies are big confinement facilities with large manure lagoons. To qualify as "organic," dairies have not been required by USDA to truly keep their cows on pasture. Rather, they only have to provide some outdoor access, use feed that has been grown without chemicals, and follow certain restrictions with respect to veterinary care. In some instances, these standards can actually end up shortening cows' lives. One organic dairy farmer I met who ran a large confinement facility in Colorado told me that if a cow gets sick he just sends her to slaughter rather than medicating her. Under the organic regulations, that's the cheapest solution. The shortcomings of the regulations have created an "organic" dairy farming that is not the answer I am looking for.

Fortunately, I believe the graziers are showing our nation a better way to produce milk. By raising cows with sound bodies and giving them a healthier lifestyle and diet, their animals live longer, have fewer calving problems, lower rates of mastitis, and fewer foot and leg problems. With lower feed costs, grazing farmers are less subject to the vagaries of the commodity markets. Because they're happier in their work, graziers are better emotionally equipped to weather the industry's ups and downs. Because their farms require less fossil fuel, both directly (by using less machinery and automation) and indirectly (by using less agricultural chemicals), grazing farmers are not as vulnerable to spiking energy costs. Their operations are also gentler on the environment. And they do us all a service by creating beautiful, bucolic landscapes.

Grazing dairy farming does require a shift in thinking. We will have to abandon our Milk Grand Prix. A corn-fed cow that gives 20,000 pounds of milk per year may give only 17,000 to 18,000 pounds a year when fed grass instead, says Ohio State professor David Zartman. Other experts estimate that the difference in annual milk production may be greater. Regardless, Zartman is a strong proponent of grass-based farming. He believes that grazing results in healthier animals with more longevity, which ends up more than compensating farmers for that loss in production.

From my perspective, it's well worth it to pay a little more for the milk, yogurts, and cheeses I buy. Whenever possible, I buy dairy products from cows on pasture. I'd rather eat a little less of it, and enjoy all of it more. As a dairy lover, I am truly grateful that such a positive alternative is entirely achievable and increasingly available. We just need to start supporting it.

CHAPTER NINE

A Bit About Fish

FACTORY FARMING THE WATERS

Before that fateful fish dinner with my family when I pronounced my intention to abandon the consumption of all animal flesh, I had been a regular eater of fish. At least twice a week, lunch had been an open-faced tuna salad sandwich on rye with pickle slices on top. And just as Mutti had always served steak on Saturdays, every Friday she cooked fish. Mutti grew up spitting distance from the Baltic Sea and loves to be near any large body of water. To this day, she has a strong gustatory and emotional attachment to fresh fish, still preferring it to any other food. Until the place closed in the late 1970s, Mutti always made a special trip to the fishmonger every Friday morning.

Yet as a child, even when I was standing alongside Mutti at the fish shop, eye-level with a line-up of whole fishes on beds of ice, I cannot remember ever thinking about how those fish had lived. I imagine I just pictured them all gaily swimming in the ocean until being harvested by small groups of burly men wearing yellow foul weather gear. At the time, that might not have been too far from the reality. In 1970, fish farming accounted for just 4 percent of world fish and seafood supplies.

However, fish farming (also known as aquaculture) has grown steadily over the past few decades. By 2000 it produced 27 percent of the world's fish and seafood. And by 2007, nearly one-half of the seafood consumed in the world came from fish farming, making aquaculture the globe's fastest growing animal food-producing industry, according to the United Nations Food and Agriculture Organization (FAO). "Worldwide, the sector has grown at an average rate of 8.8 percent per year since 1970, compared with only 1.2 percent for capture fisheries and 2.8 percent for terrestrial farmed meat production systems over the same period," the FAO has reported.

When I embarked on my exploration of the meat, egg, and dairy sectors, I did not expect it to lead me to fish. But as the details of industrial animal production became clear to me, I discovered that aquatic creatures, both farmed and wild, were actually an important part of the animal factory farming system. There are two main links: feed and waste. These connections make it important to say a few words about fish.

Fish farming is done several different ways, depending on the species raised and location of the operation. For fish species that humans have been unable to successfully breed, (tuna, most notably) the fish are corralled in the wild then kept and matured in huge nets, a process that is usually referred to as "fish ranching." Some operations raise saltwater species, often keeping the creatures in cages or nets on the coast or offshore. Others raise freshwater fish, generally in man-made ponds that resemble factory farm manure lagoons.

That's just one of many ways that fish farming mirrors industrial animal production. The industries have strikingly parallel modes of operation. Fish farming concentrates animals (and yes, fish are as much members of Kingdom *Animalia* as cattle and pigs) that would naturally roam vast expanses. Fish are placed in densely crowded containments where they constantly bump up against one another; exercise is out of the question. Their normal diet is replaced with man-made feed pellets. In short, their daily existences bear no resemblance to the lives of their wild brethren. These close quarters cause stress and easy disease transmission, leading to dependence on large amounts of chemicals and pharmaceuticals, especially antibiotics. Such concentration and extensive drug use also make fish farming waste a major pollution concern. A large salmon farm can pour as much untreated liquid waste into the sea as is generated by a small city.

To me, this all sounded eerily familiar. The similarities of aquaculture to other industrial animal operations have been noted by both fish farming critics and its supporters. An article in the *Los Angeles Times* referred to fish farms as "feedlots of the sea"; a professor of fisheries from the University of British Columbia bluntly stated, "They're like floating pig farms." Even a book for new aquaculturalists noted the similarities: "[T]echnology is now available for farming an increasing range of finfish species by intensive methods akin to broiler chicken production. As with broilers, production economics favor intensification."

Arguments made in favor of expanding fish farming echo those advanced for industrial animal production. A U.S. Department of Agriculture aquaculture report from 2004 lists the following purported benefits to fish farming: more uniformity in seafood size and quantity; selective breeding, which can improve feed conversion; and lower consumer prices. And, as with other forms of industrial animal production, aquaculture is also justified as necessary to feed the world's growing human population.

This last point (and one must allow a fish pun here) is a red herring. It presents a false choice between mass starvation and eating ever greater amounts of fish. Because this specious argument is widely made by agribusiness, it will be given special attention later in this book.

Here I mention the "feed the world" argument to point out that in the fish context it is commonly coupled with overfishing data. For example, the manual *Intensive Fish Farming* attributes the rise of aquaculture to "a combination of escalating fuel costs and declining wild fish stocks." This lays a positive environmental and humanitarian gloss on the entire fish farming industry. Aquaculture proponents are fond of touting their trade as a "blue revolution," following along the lines of the controversial "green revolution" that introduced fertilizers and hybrid seeds to developing nations in the 1960s.

That the world's wild fish populations are imperiled is well established. An alarming study by leading scientists in the journal *Nature* in November 2006 concluded that if current fishing levels and practices continue, the fisheries of all major fish species will collapse by the year 2048. "Our data highlight the societal consequences of an ongoing erosion of diversity that appears to be accelerating on a global scale," the scientists announced. "This trend is of serious concern because it projects the global collapse of all taxa currently fished by the mid–21st century." Fish farming, we are often told by the aquaculture industry, is necessary to offset this depletion of fisheries. There may be some truth in that, but the question is not nearly so simple. My research revealed that fish farming can damage aquatic ecosystems and can actually consume more fish than it produces. More on that in a moment.

Fish farming is also directly, physically connected with industrial animal production in at least two ways. First, fish are used as feed at many confinement operations. Second, several types of waste from industrial animal facilities are fed back to farmed fish. It's kind of like a two-directional pipeline shuttling feed and waste between fish farms and factory farms.

First, a look at the feed part. We've already seen that fish meal is a feed additive for the dairy industry. Valued as a protein source, it consists of fish or fish by-products that have been dried and ground into a meal. In the 1920s, fish meal was common as livestock feed in other parts of the world (notably Europe). It was being tried for the first time in the United States, primarily as protein for calves, dairy cows, and swine. But its use did not quickly become widespread. Globally, only about 8 percent of fish went into animal feeds in 1948.

In the latter half of the twentieth century, however, fish meal and fish oil became major components of farmed animal feeds. This rise coincided with both the growth of industrial animal operations and the expansion of aquaculture. A paper by Canada's Organic Seafood Association notes: "The extensive use of fish oil in animal feeds did not actually occur until the expansion of the aquaculture industry during the 1970s and 1980s."

In recent years, fish meal and oil have become major feed ingredients for farmed predatory fish, livestock, and poultry. By the late 1990s, about one-third of the global (wild) fish catch was used as animal feed. And by 2007, 37 percent of global seafood (wild *and* farmed) was ground up for use as animal feed, according to the University of British Columbia Fisheries Centre. That year, 45 percent of global fish meal and fish oil production went to the world's livestock industry, mostly to poultry and pigs.

Industrial swine operations are especially fond of fish by-products for nourishing young piglets. Traditionally (and still today on places like the Willis Free Range pig farm) pigs are weaned from their mothers when they are five to seven weeks old. However, as a cost reduction measure, industrial operations have dramatically shortened piglets' time with their mothers—weaning them as early as two weeks of age. Industrial facilities use fish meal as a substitute for the protein of mother's milk on these prematurely weaned piglets.

In my regular travels to North Carolina, I also learned that wild fish along the Atlantic coast were being pulled from the sea to feed confinement poultry and hogs, and farmed fish. Recall that the slim silvery fish I'd seen in Rick Dove's net were menhaden. The menhaden population has been struggling for years. This, in turn, is making it harder for the striped bass to survive. Journalist Dick Russell, who's written a book called *Striper Wars* describes the situation this way. "With menhaden in decline, the recovered population of striped bass aren't getting enough to eat. Emaciated stripers are being seen all along the Atlantic coast." Menhaden are in short supply, Russell explains, largely because they're being overfished by large corporations that use them to make animal feed and other products. Russell also blames the Menhaden decline on overnutrification of rivers by industrial hog and poultry operations, the very thing that Rick Dove has spent the past decade fighting to remedy.

Now, to get back to the two-way pipeline between fish farms and confined animal operations. While more fish come in the front doors of livestock and poultry facilities, stuff going out the back doors heads to fish farms. Many farmed fish are eating various waste products generated by the confinement animal industry. Poop is one favorite. Just as poop is fed to livestock, so it is being added to the feed of farmed fish. A company called Kentucky Enrichment, for instance, promotes itself as helping to solve "the problems of [chicken] waste disposal" by turning poultry waste into feed for livestock and farmed fish.

In Asia, this is often done with little or no waste treatment. "Chicken pens are frequently suspended over ponds where seafood is raised, [directly] recycling chicken waste as a food source for seafood," the *Boston Globe* reported in 2007 regarding Chinese fish farms.

Feeding poultry litter to farmed fish may even be playing a role in spreading avian influenza. A 2007 report by the Woodrow Wilson International Center for Scholars describes a suspicious incident involving thousands of migratory birds falling ill near several large carp farms and feed manufacturing facilities. "There is some speculation," the report states, "that poultry litter-based fish feed may have played a role in spreading this disease to wild bird populations."

Farmed fish are also fed wastes from slaughterhouses and rendering plants that process poultry, pigs, and cattle. The U.S. Department of Agriculture has given favorable consideration to allowing such wastes as feed as it forms standards for so-called "organically farmed" fish. "Slaughter by-products provide essential amino acids for naturally carnivorous or omnivorous finfish and crustacea (such as salmonids, marine finfish species, and shrimp)," USDA's National Organic Standards Board stated in 2006. "Use of organic slaughter by-products would also encourage waste reduction and nutrient recycling," the USDA body pointed out.

The fact that fish would never eat such foods in the wild was apparently not deemed important. But it could turn out that it is. "It is the fish farmer's nightmare that a [Mad Cow] equivalent lurks for farmed fish," says Charles Clover in his book about the state of the world's fisheries, *The End of the Line.*

Feeding slaughterhouse wastes to farmed fish is common in Asian countries, some of which import the by-products from developed nations. For example, the feed industry in Bangladesh imports large quantities of meat and bone meal from European slaughterhouses for use in fish feed. An FAO report on fish feed formulation in China lists the following animal ingredients: blood meal, cattle stomach meal, pig fat, fish oil, fish meal, feather meal, liver meal, lung meal, and meat and bone meal.

Asia and the Pacific region now account for 92 percent of the world's aquaculture, with China as the globe's leading producer and exporter of seafood. In 2004, it was responsible for an astonishing 70 percent of the world's aquaculture production.

Raising fish in ponds is an ancient Chinese tradition. "Fish farming was probably first practiced as long ago as 2000 B.C. in China," notes an aquaculture manual. Traditional Chinese fish farming was based on natural cycles. Farmers dug ponds around rice paddies and fed carp in the ponds with weeds from the rice fields. The silt from the ponds fertilized the rice fields, and crabs, not chemicals, were used to control pests.

Today, however, fish farming in China (and elsewhere) bears little resemblance to traditional, sustainable Chinese methods. China's virtually unregulated urbanization and industrialization has caused rampant contamination of air and water. A recent front page *New York Times* story characterized China's pollution problem as having "shattered all precedents." Waters used for fish farming have become so contaminated that many Chinese fish farmers douse their fish with drugs and chemicals just to keep the creatures alive.

Fish farming is a much smaller industry for the United States, ranking tenth in the world in 2004. Almost three-quarters of U.S. aquaculture is in the South (especially Mississippi, Arkansas, and Florida) and is mostly channel catfish raising done in inland ponds.

Thus, the vast majority of fish and seafood eaten by Americans is imported. Meat and dairy (except lamb) are mostly of domestic origin. In contrast, almost 80 percent of total fish and shellfish, including almost 90 percent of shrimp, come from outside the United States. For this reason alone, fish farming practices in other parts of the world matter to Americans. And China is the largest supplier of seafood to the United States.

While China's severe water pollution has rightfully triggered international scrutiny, similar fish health problems plague fish farming everywhere. "[H]igh density fish farms are natural breeding grounds for pathogens," explains Dr. Ray Hilborn, a University of Washington fisheries biologist. Partly, this is simply because water is an excellent disease transmitter. "Microbial infection and multiplication are more likely to occur from the aqueous medium," explains *Intensive Fish Farming.* "The more aquaculture there is, the more disease there will be," warns Callum Roberts, an expert in marine conservation at University of York in England.

The daunting challenge of staving off diseases at aquaculture operations has led to rampant use of drugs and chemicals throughout much of the industry. In the United States, antibiotics are commonly found in formulated fish feed. Many Asian fish farmers simply dump drugs and chemicals into fish farming waters. Shrimp is the most popular seafood in the United States and Thailand is the top U.S. supplier. Yet shrimp in Thailand and other importing countries is produced "with lax standards, raised in ponds where many chemicals are used to keep the shrimp alive." Drugs given shrimp and fish "pass easily into the surrounding environment, and some are highly toxic."

The problem of toxic compound misuse in Chinese aquaculture has lately become severe. In 2007 the U.S. Food and Drug Administration announced it would detain several types of farm-raised seafood at the border until they were proven free from residues. The FDA cited "current and continuing evidence that certain Chinese aquaculture products imported into the United States contain illegal substances that are not permitted in seafood sold in the United States." FDA sampling had repeatedly found farm-raised seafood imported from China contaminated with the antimicrobials nitrofuran, malachite green, gentian violet, and fluoroquinolone. Nitrofuran, malachite green, and gentian violet have been shown to be carcinogenic in lab animals. The use of fluoroquinolones in food animals contributes to antibiotic resistance, the FDA noted.

Parasite infestation is another chronic problem at seafood farms. "One of the most damaging organisms is the sea louse, which breeds by the millions in the vicinity of captive salmon," *Time* magazine reported in 2002. Recent research has established that lice are not only troubling for aquaculture, they are even imperiling wild fish populations in the vicinity. "Parasites that breed in fish farms kill so many passing juvenile wild salmon that they threaten the survival of fish populations in some rivers and streams," University of Alberta researchers reported in 2007.

Fish farming can also endanger wild fishes by damaging aquatic ecosystems. The FAO has been a major proponent of fish farming.

Yet its 2006 report *State of the World's Fisheries and Aquaculture* acknowledges that the aquaculture industry has caused many widespread environmental problems including nutrient pollution to surface waters; disruption of aquatic plant and animal communities; and depletion of resources (including water).

In some parts of Asia, shrimp farming practices have been especially destructive to the environment. The FAO report blames them for wetlands degradation, water pollution, and salinization of land and freshwater aquifers. In Thailand, with its 25,000 coastal shrimp farms, "long strips of coastline south of Bangkok now look like powdery gray moonscapes," according to *Time* magazine. "Shrimp farms can raise the salinity of the surrounding soil and water, poisoning the land for agriculture. Some flush their effluent into the sea, killing mangrove trees," notes the *Time* article.

Wild fish are also endangered by aquaculture's escapees. Captive fish regularly get out of their containments. "Nets are ripped open by predators or storms, fish in ponds get swept into channels by rainfall, others are released accidentally during transport." Escaped fish compete with or consume native fish, or crossbreed with them, diluting the genes that have helped wild fish survive.

Such escapes especially worry scientists monitoring wild salmon populations. Much like industrial poultry, pigs, and dairy cows, genetically altered salmon are designed for large size and quick growth but not hardiness or survivability. A scientific research team at Purdue University conducted experiments on the effect of adding larger, genetically altered fish to a wild fish population. Biologist Rick Howard and animal scientist William Muir found that genetically modified fish quickly spread their genes in the population. The genetically altered fish "has a reproductive advantage over its natural counterpart," said Howard. The scientists also found that the genetically modified offspring had poor rates of survival. "Though altering animals' genes can be good for humans in the short run, it may prove catastrophic for nature in the long run," the scientists said.

Howard and Muir were most alarmed by the combination of

the speed the modified genes spread through the population and the succeeding generations' reduced survivability. "Putting both of these things together, a population invaded by a few genetically modified individuals would become more and more transgenic, and as it did the population would get smaller and smaller," explained Howard. Over the generations, he says, the effect could decimate the wild population. "We call this the 'Trojan gene effect.'"

Salmon farming also poses other serious concerns. The raising of salmon and other predatory ("top of the food chain") fishes has been widely criticized because it requires harvesting huge amounts of wild fish as feed. In recent years, salmon has been all the rage in restaurants and grocery stores, making it now the third-most consumed seafood in the United States. Correspondingly, salmon farming has exploded, increasing forty-fold over the last two decades.

However, raising "top of the food chain" fish actually results in an overall loss of fish for the world. An analysis in the scientific journal *Nature* showed that because such fish are fed diets rich in fish meal and fish oil made from wild caught fish, the farms actually consume more pounds of fish than they produce. Specifically, "producing one pound of carnivorous farmed salmon or shrimp requires about three pounds of wild fish in the form of fish meal," said Rebecca Goldburg, the study's lead author. Similarly, ranched tuna, which dine on live pelagic fish, such as anchovies, sardines, and mackerel, take about twenty pounds of feed for every pound of tuna meat.

There is also good reason for concern about the safety of eating farm-raised salmon. Several studies have found elevated levels of various toxins in farmed salmon. These higher toxin levels occur in farmed salmon because the contaminants are concentrated in salmon food, which may be made from small fish from contaminated waters. In a major study in 2004, researchers found that farm-raised Atlantic salmon from both Europe and North America had significantly higher levels of thirteen toxins (including PCBs and dioxins) compared with wild Pacific salmon.

Yet another unsettling fact about farmed salmon is that they are fed red dye. The diet of farmed salmon lacks the small, pink-colored krill they would eat in the wild. Without krill, their flesh is gray. Just as the dye cantaxanthin is added to the feed of confined laying hens to make their egg yolks yellow, so cantaxanthin is added to the feed of salmon to make their flesh pink.

Clearly, simply continuing to expand fish farming, especially of predatory fish, is no solution to the world's complicated fish problems. In *The End of the Line*, Charles Clover concludes that aggressive management of wild fisheries is much more important. He also argues, and I agree, that the wealthier global consumers (especially in Japan, Europe, and the United States) should influence fish catching and raising through more informed and restrained choices. He writes: "[I]t is the responsibility of those of us who are fortunate enough to exercise choice to ask whether this is the way the world should be going." Restaurants, where 70 percent of U.S. seafood consumption takes place, will have to be at the forefront of the solution.

Part of this responsibility also includes re-evaluating the wisdom of using fish, both farmed and wild, as livestock and fish feeds. In particular, this practice makes little sense when used to enable questionable practices like premature weaning.

I'm glad that, as a person who no longer eats fish, I don't have to struggle with the complex question of how to select fish as a consumer. Given the state of the wild fisheries and the problems with fish farming, it seems that there are no good choices. If I were a fish consumer, knowing what I now know, I would choose primarily wild, line caught fish; and I would limit my overall consumption of fish to once a week or so.

Finally, I'd like to say a word about fish as animals. When I was ten years old, several goldfish were given to me. They lived together in a wide shallow glass bowl with some sea glass and ceramic figurines. I often sat near it waiting for indications of what was going on in their universe, such as it was. It was easy to see that each fish had an individual character, and I named each according to its behavior and appearance. One was placid while another was jittery. The third was quite a bully. At least that's how I viewed them at the time. Otherwise, their unblinking saucer eyes and implacable faces made it hard to detect personalities and emotions.

I later learned that scientists have in fact long documented the existence of fish personalities. A 1915 article in the *New York Times* describes the work of Dr. Francis Ward, an English zoologist, whose experiments showed fish displaying signs of fear, suspicion, and doubt. "We are accustomed to think that only we humans become pallid with fear or agitated with joy," said Dr. Ward. "But some experiments . . . with perch show that when their repose is suddenly disturbed with tapping on the glass the fish visibly tremble and the [stripes] which are characteristic of this species actually disappear for the time being, only to reappear when the disturbance is removed and the equanimity of the fish is restored."

More recent lab work has reached similar conclusions. In 2006, scientists at the University of Liverpool conducted experiments showing fish altering their behavior after watching other fish. The researchers presented the fish with Lego blocks to elicit a fear response. They then observed that bolder fish approached the object within just a few minutes whereas shy fish took more than ten minutes. "[T]he more time bold and shy fish spent watching each other the more their behavior changed," the scientists noticed. "Bold fish who observed the way shy fish reacted to the Lego objects became much more cautious in their behavior. In contrast however, shy fish who observed bold fish did not alter their behavior—they remained just as shy as before." A 2007 study by researchers at University of Guelph likewise concluded that fish have distinct, discernable personalities.

Moreover, the fact that fish have feelings has even been tacitly acknowledged by the aquaculture industry itself. Stress is an emotion and the industry has recognized that certain farming conditions cause the fish stress. "If a particular species of fish becomes stressed when crowded under farm conditions, it is likely to stop feeding and consequently growth and survival rates will be poor," states the industry guide book *Intensive Fish Farming*. So it's widely accepted that fish experience negative emotions.

But what about positive emotions? I've never come across a study documenting fish happiness. A wonderful book about animals' positive feelings, *Pleasurable Kingdom*, points out that when it comes to fish, "there is practically no research in this area (yet)." However, the author goes on to note that fish almost certainly do feel pleasure and happiness. He argues that negative and positive emotions serve equally useful evolutionary purposes. "Just as pain helps an animal that can detect and escape aversive stimuli, pleasure is useful to an animal with a sophisticated nervous system, cognitive skills like learning, long-term memory and individual recognition, and which can discern, seek out and locate important rewards." Therefore, he reasons, it's just as likely that fish feel positive emotions as negative ones.

This makes complete sense to me. My guess is that fish feel especially happy when they're at play, when they're reunited with a schoolmate, and (except for the bottom-dwellers) when soaring through an expanse of cool, crystal clear water.

I bring up the emotional lives of fish to reinforce the point that they are sentient living creatures, a fact that seems often forgotten (or perhaps is not widely known). In whatever way we raise animals for food—including fish—we should ensure that they do not suffer unnecessarily. Having never visited a fish farm, I don't yet have a clear picture of what a good one would look like. But I do believe that, like all forms of animal farming, along with proper feeding and waste management practices, it should give consideration to the quality of life of the creatures it is raising.

CHAPTER TEN

Finding the Right Foods

THE MAKING OF A FOOD DETECTIVE

Eating the right foods is an important part of my daily life. Over the years that I've been learning about farming's industrialization, I've turned into a veritable food detective in seeking out foods from traditional farms. I just feel better when I'm eating foods that were produced in ways I'm comfortable with. And I believe that traditionally farmed foods are better for my health. I know they taste better.

Mutti and Vati taught me early the importance of paying attention to my diet. Before I was born, after my father's parents had both succumbed to heart disease, he became obsessively attentive to every morsel of food he put into his mouth. He considered exercise and eating right the cornerstones of good health and longevity. Our family meals invariably included ample fruits and vegetables (many of which we had harvested ourselves). Grains were whole:

rice and bread were brown. Beverages were tap water, juice, or milk (never soda), with an occasional glass of beer or wine for the adults. Sugar, salt, and fat were consciously kept to a minimum. Every dinner included a large salad with Mutti's house-made dressing—a simple mixture of oil, vinegar, salt, and pepper. Except on special occasions and weekends, dessert was fruit—whatever was fresh and in season or, during the winter, a whole, peeled orange or grape-fruit, or a baked apple stuffed with raisins.

That may all sound unappetizingly ascetic but my parents had no interest in suffering through their food and never treated eating what's good for you as a chore. Meals, they believed, were among life's greatest pleasures and should be full of delicious food and lively conversations.

Mutti also taught my siblings and me a lot about preparing food. Since she made almost everything from scratch, she often enlisted us as kitchen help. As we matured, we were involved in all stages of meal preparation and received instruction in the art of bread baking. By the time I headed off for college, I had kitchen basics pretty well down.

Mutti's lessons were later supplemented by outside influences and my true food passion blossomed. During college I earned pocket money with waitressing stints in three restaurants, where I picked up additional cooking techniques. Once I graduated from law school, I spent a lot of time in my own kitchen, collecting cookbooks, and soon becoming addicted to *Cook's Illustrated* magazine, which offered recipes and sound culinary advice. Like Mutti, the good folks at *Cook's Illustrated* repeatedly emphasized that the foundation of the best cooking and baking is always first rate ingredients.

So when I started my Waterkeeper job, I had decades of kitchen experience under my belt. But my long work hours and pint-sized Manhattan kitchenette nearly ground my culinary exploits to a halt. I was practically living on take-out, often pecking at it while work-ing at my desk. Even so, I stocked my cupboards and fridge with the basics and composed simple meals at home whenever possible.

Grocery shopping, however, was becoming complicated. The more familiarity I gained with how animals in our food system are treated, the more I felt the need to supply my kitchen in a way that was in keeping with my values. I was developing an aversion to eating foods from industrial facilities, and I hated the thought of my consumer dollars adding to their profits. But how could it be avoided? The products of industrial agriculture are ubiquitous and often seem like the only option. And even if I could find stuff that *appeared* to be from traditional farms, how would I know which claims and labels I could trust? This was the challenge.

Eventually, I realized that the solution to navigating the consumer food maze lay in returning to the kind of food foraging in which my mother had always engaged—gathering foods from a variety of places, getting as close as possible to the original sources. I'd have to get a little creative and would need to venture beyond the well-worn rut from my apartment to my habitual supermarket. Reluctantly, I also came to accept that I couldn't achieve humane, sustainable shopping perfection overnight. It would be a transition over time in which I was continually seeking more information and looking for ways to improve. ("Baby steps are okay, as long as they're in the right direction," a friend of mine wisely said.)

THE EGG HUNT

Take eggs, for example. These days I eat the most exquisite eggs I've ever encountered. They come from a flock of vigorous hens of traditional, colorful breeds. The hens have the run of a huge grassy yard with trees and other places to roost, laying their eggs in sheltered, straw filled nests. The woman who keeps this flock knows the birds individually and would never dream of clipping their beaks. She takes tremendous pride in her fowl and the beautiful eggs of many shades they produce. Finding such a source took patience and persistence.

I made my first moves away from industrial eggs while living in Manhattan, inspecting both the "cage free" and "organic" varieties. "Cage free" hens are not kept in the small, wire mesh boxes (battery cages) that are standard in today's egg industry. That's a good thing. But my research revealed that it's just one step in the right direction. "Cage free" operations are not required to provide hens access to the outdoors and most do not. The facilities tend to be huge, extremely crowded industrial contract growing operations.

On the other hand, *organic,* a word regulated by USDA, holds a lot more meaning. To be labeled "organic" eggs must come from hens provided only organic feed (which has at least 80 percent USDA certified organic ingredients and does *not* contain slaughterhouse wastes, antibiotics, or genetically modified grains). The organic standards also provide some assurance about how the animals are housed and handled. They require that organic livestock (including poultry) be provided "living conditions which accommodate the health and natural behavior of animals," and specifically mandate that animals have some access to the outdoors, to exercise, and to bedding.

There's a lot of wiggle room in interpreting those terms and, unfortunately, some companies do the bare minimum. The outdoor space provided laying hens may be small cement or gravel areas, not roomy pastures. Still, the organic standards do require a higher standard of feed and better animal living conditions for hens than ordinary grocery store eggs (for which there is essentially no standard). Buying only eggs at my supermarket that were both "organic" and "cage free" seemed a couple of baby steps in the right direction.

However, I wasn't really satisfied—I wanted to actually know how and where the hens were raised. The egg company I was buying from had a folksy name and a picture of two guys wearing overalls and straw hats. It was clever marketing but didn't allay my concerns about whether the eggs came from a place that I would feel good about if I visited. I decided to dig a little deeper. I found an Internet address on the carton and went to the company's website. Unfortunately, it didn't tell me much. There was nothing specific about the hen feed and neither photographs of the hens nor of the places they were housed. Don't give me photos of the products or drawings of Old McDonald's Farm—*I want to see actual photos of the animals and places my food comes from!*

A couple of months after I started buying this brand I became even less satisfied with the eggs when they started tasting fishy. My suspicion was the hens were being fed hefty doses of fish meal. At that point, I made up my mind to write the company inquiring what was in the feed. A simple enough request, one would think. We've all been taught that we are what we eat and, obviously, if we eat animal products, *we are what they eat, too.* But oddly, the United States has no law requiring that agribusiness companies disclose feed ingredients to the public even though residues can be found in the animals' flesh, eggs, and milk. Organic companies (at least theoretically) adhere to a higher standard, so I hoped this company would respond. My inquiry, however, got no reply. A few weeks later, I sent a second note, restating my concern over the feed (and mentioning my disappointment that I'd gotten no answer to the first letter). Still, nothing.

Even though it was the only "cage free" and "organic" choice at my grocery store, I then swore off that brand and never bought from them again. If a company won't show and tell me how they're raising their animals, I'm not interested in purchasing their food products. Until I could find a better alternative I would go eggless.

This experience was teaching me that I would need to venture beyond the one-stop supermarket. With my tight schedule, that seemed like a burden. Once I actually started exploring the alternatives, however, hunting for good food became good fun. My first foray was to a smaller, specialty grocery store a few blocks beyond my supermarket. There I found several brands of eggs bearing special claims.

One carton was labeled "Certified Humane." From my work, I

already knew that label didn't hold much promise. I'd read through their standards and discovered that industrial confinement operations could meet them. In fact, the program struck me as *designed for* big agribusiness. For example, it didn't ban metal crates for sows, failed to require that animals had access to the outdoors, and failed to prohibit systems using liquefied manure.

On top of that, I'd reached the conclusion that the Certified Humane label lacked credibility because of how it was financed— paid for on a per animal basis by the very operations being certified. Thus, the Certified Humane program has a built-in incentive to court and retain large facilities. In my view, because it heavily depends on revenue from the animal operations themselves, particularly large ones, it cannot be objective. The bottom line for me is that the Certified Humane program does little or nothing to get us any farther away from industrialized agriculture.

A far superior program, Animal Welfare Approved (AWA), has been developed by Animal Welfare Institute (in other words, by the Halverson sisters). The specific standards are available to the public at www.animalwelfareapproved.org. The program's underlying premises are that industrialized agriculture is inherently hostile to animals and that only nonindustrialized farming can be truly humane. Their animal treatment standards are the most stringent I've seen in this country. Yet at the same time, because Diane and Marlene have spent a lifetime around traditional farmers, the standards reflect the practical realities of what it takes to run a ranch or farm. Moreover, AWA certification requires that animals be raised on real family farms and explicitly prohibits industrial practices like sow crates and liquefied manure systems. Thus, the Animal Welfare Approved label cannot be earned by an industrial operation. AWA also maintains a truly independent position because it's not financed by the certified farms. Unfortunately, the AWA labeling program, although it's growing quickly, is still in its infancy. There's not much on the shelves yet.

While living in New York I couldn't get my hands on Animal Welfare Approved eggs so I ultimately settled on a brand at the gourmet foods market with the triple moniker "cage free," "organic," and "free-range." It probably wasn't much better than the brand with the two guys on the label but at least the eggs didn't taste like fish. Like other consumer claims, the term "free-range" is imperfect. USDA has done little to define it, requiring only that poultry raised for meat have access to the outdoors (although not pasture) while leaving the term wholly undefined for laying hens and other animals. Thus, arguably, egg companies can legally claim to be "free-range" even without providing hens any outdoor access, and apparently some do just that.

Sigh. I found the whole scramble of egg carton terminology somewhat discouraging. Although I felt I'd improved my grocery shopping, I doubted I was getting what I was really looking for—eggs from traditional farms. Then I moved to California and things really started looking up. Living in a rural community like Bolinas was a real advantage because it actually has a poultry population. In our town, my guesstimate is that a couple dozen families keep backyard flocks. Such households are often thrilled to sell their extras and are the most fertile sources of the very best eggs.

Originally, I planned to immediately start a hen flock when I arrived in Bolinas. But I soon realized that putting the idea into practice would require careful planning because of the robust populations of rats, raccoons, raptors, foxes, bobcats, and coyotes who make our ranch their home. Until I devised a sound protection plan for pastured hens, I would seek out a local source of eggs. Our community food co-op carried eggs from a nearby family farm that I started buying. But because I could see that the hens were not given access to true pasture, my quest continued.

Several months later, I came across a ranch in our area with a temporary white sign announcing: "Eggs for Sale." Instantly, I U-turned, pulled into the driveway and practically ran to the front door. A kindly man answered my knock. I asked him several questions about how the hens were raised and what they were being fed. Vegetable scraps and organic grains, he said, and told me where he purchased the feed. Then I asked if he would mind showing me the hens and their living quarters. The man raised his eyebrows slightly and paused. After a moment, he slowly answered: "Well, no one's ever asked me that before, but I'm happy to show you."

I followed him to a small wooden building with clean wood shavings covering the floor. Straw-filled nesting boxes and perches were mounted on the walls. A wide doorway opened into an attached yard of grass and wildflowers. I stepped through the opening. About twenty hens of black, red, and variegated black and white were busily pecking at the ground, preening themselves, chasing one another, and basking in the sun. I turned to the man and nodded. "Perfect. I'll take whatever you can spare."

When that ranch was short on eggs for a while, I located yet another person in town with a poultry flock that ranges in pasture and eats only natural feeds. I'd heard about her through word of mouth. The homeowner and I reached an arrangement that I'd stop by for eggs every Thursday. She keeps five or six chicken breeds, each providing a distinctly different egg. In color, the eggs are everything from pale blue and green to a flecked dark terra cotta and deep tan. In size, they vary from small to jumbo. Every dozen is an assortment. I look forward to opening a carton of these eggs almost as much as a box of chocolates.

As lovely as they are to look at, the greatest reward is in the eating. The shells are solid and don't burst when I boil an egg for my breakfast. Other mornings, I make Bill and myself fried eggs, one of our favorite meals. When I crack open one of these eggs, the whites are thick, never runny. The yolks are dense and deep golden, like a ripe persimmon. The color's intensity comes from the xanthophylls of the plants on which the hens munch. I'm sure their happy lives help, too. The mass produced variety—even if they're "organic"—simply cannot hold a candle to the richness, flavor, and beauty of these eggs. I still dream of keeping our own flock, but for now I'm utterly content with this alternative.

WHERE TO FIND FOODS
FROM TRADITIONAL FARMS

My search for nonindustrial eggs taught me quite a lot. Most important, I realized I had to stop being a supermarket zombie. Finding foods from traditional farms and ranches involves being something of an explorer and a detective—being willing to go places that are off the beaten path and to ask questions about where food is from and how it was produced. Mainstream supermarket chains are carrying more organic foods these days but are still mostly arid wastelands for sourcing foods from traditional farms and ranches. I've dropped comment cards into supermarket suggestion boxes over the years and I figure the more people they hear from the better, but there's a long way to go.

A few steps ahead of the rest is Trader Joe's. In our area, Trader Joe's has long carried meats from Niman Ranch and organic dairy products from the Straus Family Creamery. These brands are supplied by traditional, pasture-based farmers and ranchers who don't use hormones or feed antibiotics. Both have consumer friendly websites with specific facts and photos showing and telling how their animals are reared and fed.

More fruitful than conventional grocery stores are natural food co-ops. Although in the past many allowed only members to shop, these days most co-ops welcome the public. Our local co-op sells eggs and dairy products from nearby farms that we're familiar with. I've been in co-ops in many parts of the country and have often found them to carry the products of smaller scale, traditional farms. They're usually staffed by people knowledgeable and passionate about natural food who are only too happy to answer questions. Co-ops pride themselves in responding to consumer concerns and therefore are likely to start carrying products from pasture-based farms if customers ask for them. The co-ops in Boise, Idaho; Bozeman, Montana; and "The Wedge" in Minneapolis are not to be missed. Since I've never found a national directory of co-ops, when I'm looking for a co-op in an unfamiliar town, I just ask around.

Independently owned specialty food stores are another good bet. They compete with big-box supermarkets largely by carrying products not offered at national chains, often including the products of local farms. And, even more so than co-ops, they tend to be extremely responsive to customers, which makes them a great place to request things like eggs and pork from animals raised on pasture. San Francisco has Bi-Rite Market; Denver has Marczyk Fine Foods; Des Moines has Gateway Market; Brevard, North Carolina has Poppies Gourmet Farmers Market. Just about every urban center has one or more stores of a similar ilk, which are usually listed in the phone book under something like "Gourmet Food Shops & Specialties."

Like my mother always did, I now buy directly from farmers whenever possible. From May through October, at least once a week, Mutti headed to farms with roadside stalls that featured their daily harvest. Other than what she grew in her own garden, it was the freshest produce my mother could find. Fortunately, Bolinas is blessed with a wealth of organic family farmers. Getting to know them and their farm stands has been a pleasure. Often what I'm buying was picked that morning. The produce is colorful, fresh, and flavorful—truly a joy to prepare and to eat.

Prince Charles, himself up to his knickers in organic farming, visited Bolinas in 2005 just to tour several of our local farms. For about eight months of the year, I get most of my vegetables, greens, herbs, honey, and cut flowers from the farms visited by His Royal Highness, all of which are within five miles of our ranch. I get a kick out of knowing that these farms were recently graced by the presence of British royalty. But what pleases me more is that one of them is the venture of a man in his twenties full of energy and passion for organic farming. The farm stands in this area, as many others, are limited to fruits and vegetables. But some also offer eggs, poultry, or even meat. Tracking down a community's farm stands is like hunting for buried treasure—it takes some effort but the rewards are great.

Mutti also purchased directly from small farmers who brought their wares to town once a week (as with our eggs). When she finally became fed up with the bland, flaccid chicken offered at the supermarket she began seeking something better. Then she discovered Otto, a local Amish farmer who sells whole chickens and homemade chicken sausage from a van that shows up at a particular parking lot in town on a certain day every week. The meat was darker, succulent, and full of flavor. "*Ah*, now that's what chicken is supposed to taste like," my mother crowed over Otto's chicken. Mutti paid more per pound for it and offset that by cutting back a bit on our portion sizes. No one in the family protested.

Farmers markets are another avenue for buying directly from farmers. Their explosion in popularity in the past decade has made them almost as much social scenes as shopping venues. I regularly shopped the farmers markets of Raleigh, Kalamazoo, and Ann Arbor. The weekly outing became a beloved ritual and my main source of fresh fruits and vegetables for almost half the year. These days, I have the choice of three, including a small one in nearby Point Reyes Station and the large, glamorous Ferry Plaza Farmers Market in San Francisco. Along with gorgeous produce and flowers, farmers in our area sell eggs from hens on pasture, grass fed beef, smoked wild caught salmon, and cheese and yogurt from pastured cows. Farmers markets can be located by consulting the USDA website, which maintains a list at http://apps.ams.usda.gov/FarmersMarkets.

The great advantage to buying at farm stands and farmers markets is that one can (and should) directly question farmers about how their animals are raised and fed. I've never encountered a farmer vendor unwilling or unable to answer such questions. I've also found them open to requests about what they should be offering. Chatting with farmers is an excellent way to learn about your food, even ideas for preparing it.

Another way to purchase direct is through "community supported agriculture" (CSAs), farms that offer consumers the chance to buy "shares" that entitle "shareholders" to a portion of what they produce. Several friends of mine have had positive experiences with CSAs. The food, they said, was fresh, delicious, horizon-expanding (because they received things they'd never tried before), and cheaper than comparable foods at stores. CSAs are increasingly available for eggs, meats, and dairy products. They can be located on the web at: www.eatwellguide.org and www.localharvest.org/csa.

No matter where I buy animal products, I always read the labels and ask questions. I also follow some basic rules of thumb. As a starting point (as for fruits and vegetables), I buy only products of the United States unless I know they're from a place that has similar (or better) health and safety regulations (such as countries in the European Union). Recent experiences with Chinese food imports highlight that the U.S. government does precious little to ensure the safety and healthfulness of imported food. Although Whole Foods and other large supermarket chains are increasingly sourcing their "organic" foods from China, to me, something from China labeled "certified organic" is a joke.

Knowing the national origin is especially important when buying fish and seafood, the majority of which is imported from parts of the world with lax safety regulations. Just as with other animal products, the key is asking questions about the food one is buying. Given the rampant use of drugs and chemicals in aquaculture, I would avoid buying imported seafood unless it came from Canada or the European Union.

Untangling the many complicated questions surrounding seafood purchasing requires calling in the experts. Monterey Bay Aquarium has published a handy wallet-sized guide called "Seafood Watch," useful for knowing which fish species are abundant versus those that are over-fished. It's available at www.mbayaq.org. The guide recommends giving up bluefin tuna and farmed salmon while encouraging consumption of farmed catfish and wild caught

Alaskan salmon. Chefs Collaborative, an organization promoting environmentally sustainable purchasing to chefs (and on whose board I sit) completed a useful seafood buying guide in 2008, which is available on the Internet at www.chefscollaborative.org. A couple of private sector companies, one called CleanFish (www.cleanfish.com) and another called Ecofish (www.ecofish.com), have developed credible consumer labels to identify fish farmed or caught using sustainable practices. Here in Bolinas, it's even possible to buy line-caught fish from local fishermen. When you can get it, that's the best option as long it came from safe waters.

For me the most important thing is finding products from traditional farms with animals living on pasture. All farm animals— turkeys, chickens (including laying hens), pigs, dairy cows, beef cattle, goats, sheep—benefit immeasurably in health and quality of life from spending time on pasture. If a farm or ranch provides this, the animals probably have a reasonably natural life. For grazing animals, especially, this is arguably a *sine qui non* for good health and welfare. Eggs, meat, and dairy products from animals on pasture are better for you and more flavorful. Moreover, it's rare that a dairy, pig farm, or poultry farm that raises animals on pasture uses hormones or adds antibiotics to feed. Since USDA does not regulate the term "free-range" and the organic standards do not require true pasture, the only way to be sure the animals live on pasture is to know the farm or to ask.

The one exception I'd make to the need for pasture is for pigs raised in deep straw bedding. Farmers in Sweden and the United States have pioneered rearing pigs in spacious indoor areas (without pens) in which pigs live on a foot or more of straw. Fresh straw is regularly added to keep the pigs on clean, dry bedding. (I've seen several of these farms.) The deep bedding method mitigates the environmental impacts of confinement systems and the welfare of the animals is high, making it an acceptable alternative.

A word here about meat labeled "natural." In 2007, Bill and I participated in a public hearing in Denver, Colorado, where USDA accepted comments about the use of the term. We argued that for meat to be considered "natural," animals must have lives with some connection to how they'd live in nature. Clearly, people assume meat labeled "natural" was *raised* in a natural way, so our point was that the label should meet consumer expectations. Obviously, this would rule out meat from animals raised in industrial confinement facilities. Regrettably, however, at this time the word has no real meaning. USDA currently requires only that the meat labeled "natural" be "minimally processed." Astonishingly, *how the animal was raised is given no consideration*!

Thus, the word *natural* is shamefully abused in the meat industry. Smithfield Food, Inc. labels its pork—which comes from pigs fed questionable feeds and raised in giant metal and concrete warehouses—"natural pork." Brandt Beef (a sizeable national brand whose motto is "the True Natural") is supplied entirely by male Holstein calves taken from their mothers at birth then raised in feedlots (not on pastures) and calls this "natural beef." These examples illustrate the importance of asking questions about how animals are actually raised rather than relying on labels or marketing claims.

If I cannot be sure animals were raised on pasture, I look for the organic label. When I'm buying directly from farmers, I ask specific questions about their animal husbandry. Preferably, I see the farm or ranch for myself—at least in photographs. Or it may be a brand or organization that is a collection of farmers that follows a specified set of protocols that I trust. Under those circumstances, I'm not as concerned about whether it's certified organic. However, when I buy food at stores from brands I'm not familiar with, the organic label is a good fallback. It's not ideal but it's a solid indicator that feed and husbandry standards are high.

Things get a little dicey when none of the above is available. Then I look for indications that animals at least have outdoor access. For chicken and turkey meat, "free-range" gives some assurance. For other foods, if it doesn't say "raised outdoors" or words to that effect, it probably wasn't.

If outdoor-raised isn't available, I seek as the barest minimum: foods from animals never given hormones (such as rBST for dairy cows), never fed antibiotics, and not housed in cages or crates. Plenty of pork and poultry in supermarkets these days is labeled "antibiotic free," but most of it comes from large industrial operations. If at least all those basic standards aren't met, I just scratch the item off my grocery list.

It might be noted that I haven't included "buying local" in this hierarchy of purchasing principles. Buying locally produced foods is desirable for many reasons. The foods are usually fresher; buying them supports the local farm community and minimizes pollution from transporting food long distances. I attempt to buy all my fruits, vegetables, and flowers from local farms and ranches. Nonetheless, when it comes to purchasing foods from animals, local is nowhere near my top priority. I consider it much more important *how* the animals were raised than *where*.

A friend in Minnesota told me about a funny experience that illustrates how people can go astray with the idea of "local meat." Her natural foods co-op was carrying pork promoted as coming from a local farmer. When my friend learned that it was actually a confinement operation with liquefied manure, gestation crates, and antibiotics, she brought the facts to the co-op's attention. Amazingly, the co-op said it wanted to continue offering the pork because it was "local."

I also don't really believe in "local meat" because some regions of the country are much more environmentally appropriate areas than others for raising certain species of livestock. For example, many parts of the far West can support grazing animals but have insufficient topsoil, heat, and rainfall to raise a sufficient quantity of crops as feed for poultry and pork. Those regions are great country for cattle, sheep, and goats but poor places for animals that need to be fed grains. It's wildly inefficient (and environmentally damaging) to ship massive quantities of feed from one region of the country to another. (Shipping feed long distances can actually make the total "food miles" greater for "local" animal products.)

This is why pigs have historically been concentrated in Iowa while cattle are raised in Montana. Thus, the idea of "local food" must always be balanced with the concern for, as Bill puts it, "appropriate locale." A sustainable food system must be based on farming that's truly environmentally sound.

EATING OUT

As much as Americans are unaccustomed to asking questions about the source of food at the grocery store, I think we're even less used to making such inquiries at restaurants. In my three waitressing jobs, I'm pretty sure I was never asked about the sources of ingredients, just how dishes were prepared. Fortunately, this is changing. Famous restaurants like Berkeley's Chez Panisse and New York's Blue Hill have made it an integral part of the experience that diners know precisely where—you could probably get the GPS coordinates—their food is coming from.

Whenever Bill and I eat out, he asks the following simple question before ordering meat or fish: "Where is it from?" If the server doesn't know the answer, Bill requests that he or she go ask the chef. If the chef doesn't know, Bill doesn't order it. We're finding that more and more places are designating on the menu where their meat and fish comes from and are making sure their staff is prepared to talk about the origins of meal ingredients.

Certain chefs and restaurant owners take special pride in the quality of the food they're offering. Those are often the places that source ingredients from traditional farms and sustainable fisheries. Some of the best restaurants in the country are at the forefront of this growing movement. Just a few examples are: Lumière near Boston; Savoy and Green Table in New York; North Pond in Chicago; Highlands Bar and Grill in Birmingham, Alabama; and Oliveto in Oakland, California. Thankfully, there are many, many more. Some of my favorites, like Cactus Taqueria in Berkeley,

White Dog Café in Philadelphia, and Zingermann's in Ann Arbor, are entirely affordable.

These days it's even possible to find fast-food restaurants supplied by traditional farms and ranches. The real pioneer was Chipotle Mexican Grill. Years ago, the company's CEO and founder, Steve Ells, was looking to improve Chipotle's pork carnitas. Trained as a chef at the Culinary Institute of America, Steve was finding industrial pork to have abysmal eating quality. Then he read an article about Niman Ranch and called Bill Niman. Within about a year, Niman Ranch became Chipotle's pork supplier. Steve Ells and other people at Chipotle became so enthusiastic about the Niman Ranch farmers that they decided to begin converting the entire Chipotle menu to organic and nonindustrial ingredients, an ongoing program called "Food with Integrity." Other chains, like Pain Quotidien, and Panera, are imitating Chipotle by trying to upgrade their ingredients in a similar fashion.

The most important thing I've learned about finding animal products from traditional farms is not to get easily discouraged. There's more of it out there than one realizes, but most of it is not available in mainstream supermarkets and restaurants—not yet. Finding it requires some persistence and adventurousness but can also be fun. I'm convinced that every time a grocery store or restaurant gets a request for pasture-raised pork it's adding to a groundswell of demand that will eventually create the supply.

CHAPTER ELEVEN

Answering Obstacles
to Reform

THE INEVITABILITY
AND EFFICIENCY MYTHS

It's now eight years that I have been working on changing how animals in our food system are raised. I've found that almost all people share an interest in eating safe, wholesome food and a belief that animals should be treated humanely. A recent survey by Michigan State University confirms that most people feel those aims are best achieved by raising animals on traditional farms, finding that 80 percent of people consider meat from animals raised on pasture healthier and 96 percent consider pasture farming more humane than confinement systems.

Yet I've also realized that many people assume industrial farming is the only realistic option for producing food these days; they are resigned to it as a necessary evil. I've often witnessed someone nod sympathetically then suggest that industrialization must be the unfortunate but inevitable transition from obsolete, inefficient production methods to modern, efficient systems. Indeed, the transition from traditional family farms to industrialized agribusiness has long been treated as unavoidable (and even desirable) by our own U.S. Department of Agriculture. USDA acts as though industrialization is economic Darwinism—only the fittest forms of farming are surviving. Inevitability has now become widely accepted. But the inevitability of industrial animal production is a myth. It's not inherently more economically efficient than traditional farming and nothing is unavoidable about it.

Larger farms are not more efficient than smaller ones. Dr. Willis Peterson, an agricultural economics professor at University of Minnesota, debunked this commonly held misconception with a detailed study of the relationship between farm size and efficiency. His research concluded that size did not determine efficiency: "[S]mall family and part-time farms are at least as efficient as larger commercial operations. In fact, there is evidence of diseconomies of scale as farm size increases."

More specifically, much empirical research has shown that smaller scale livestock farms can be as efficient as larger operations, or even more efficient. For *Pigs, Profits and Rural Communities*, a book about the societal impacts of industrial hog operations, the authors reviewed all available research on economic efficiency in hog farming. They concluded that "size has little to do with efficiency of production and profitability." For example, a study of over 700 hog farms in Illinois found that "size of operations explained less than 5 percent of the variation in profitability." Research done by Kansas State University examined the records of ninety-one Kansas hog operations and found that smaller farms (those having approximately seventy-five sows) had the lowest total cost of production.

Other research has demonstrated that farms raising animals on pasture can produce food more cost effectively than confinement operations. Iowa State University, the nation's ultimate authority on swine husbandry, conducted a study directly comparing confinement hog operations with free-range farms. It determined that the total production costs to raise pigs to a market weight of 250 pounds was $4.88 *less* for the outdoor raised pigs. Overall, fixed costs were 30 to 40 percent lower and total costs were 5 to 10 percent lower for the free-range farms.

Likewise, numerous studies have found that grazing dairies produce milk at a lower cost than confinement dairies. For example, a University of Wisconsin study found that the higher costs of machinery, production, and feed for confinement systems made the pasture farms more economically viable and more profitable for farmers. This research further confirms the experiences of the Wisconsin grazing farmers Bill and I first befriended on our honeymoon, who each told us they were faring better financially with grazing. Similarly, a University of Vermont study found that grazing dairy farmers made more money per animal because they spent less on equipment repairs, fertilizers, pesticides, and fossil fuels.

Dr. William Weida, an economics professor who retired from Colorado College, closely reviewed the economics of the farm animal industry over a number of years. I first met Dr. Weida when working for Waterkeeper and have spoken with him many times over the years. He told me that the popular misconception that industrial animal operations are efficient just drove him nuts. His research demonstrates just the opposite: large industrial animal operations are economically inefficient. Summarizing his analyses, Dr. Weida has written: "[I]f all the economic costs of [industrial operations] are considered, two economic concepts—diseconomies of scale and diminishing marginal returns—both mandate that the efficient size of most animal feeding operations should be relatively small."

Diseconomies of scale, Dr. Weida explains, occur when the problems associated with production rise faster than their increased output, making operations less profitable as production grows. Weida determined that for industrial animal operations, because of the crowded, confined living conditions, disease control is such a concern. The larger the operation, the more serious the disease control problem.

Dr. Weida's research revealed that as industrial animal operations get larger they also have diminishing returns. Here we once again get back to poop. Obviously, the more animals there are at a facility, the larger the volume of feces and urine. At the point when adjacent land can no longer absorb the waste, it must be exported from the area. Weida discovered that when the operation reaches the size at which manure hauling is necessary, the facility's costs soar.

USDA documents confirm this, noting: "Animal manure is costly to move relative to its nutrient value, limiting the area to which it can be economically applied." For some operations, the closest land available for manure disposal may be many miles away and hauling costs increase with every additional mile. USDA research also shows that the larger the operation, the more expensive it becomes to transport and dispose of manure. For these reasons, per animal manure handling costs actually go up as operations get bigger.

On our Midwest Whistle Stop Tour, Rick Dove and I spent a couple of hours in Ames at the office of Iowa State University agricultural economist Dr. Mike Duffy. I was aware that Dr. Duffy had spent years studying agricultural efficiency so I was especially interested in what he had to say on the subject. I asked him whether he had concluded that industrial animal operations were more efficient than traditional farms. "Oh God, no," Duffy responded emphatically. "It's not about efficiency at all. It's about *power.*" He went on to explain that in both the marketplace and politics agribusiness has much greater power than family farmers and that this is the real reason for the dominance of industrialized animal production.

Indeed, over the past century, each major animal food sector has become controlled by large agribusiness. Today, about 86 percent of the beef industry, 64 percent of the pork industry, 56 percent of the meat chicken industry, and 51 percent of the turkey industry are controlled by the top four agribusiness giants of the respective sectors. About 60 percent of eggs are produced by vertically integrated corporations and almost all of the rest are produced for those corporations under contract. Dr. Neil Harl, another Iowa State University agricultural economist I've met, says that the effect of such monopolies has been to turn family farmers into "serfs." "A producer without meaningful competitive options is a relatively powerless pawn in the production process," Harl says.

Clearly, then, greater economic efficiency does not explain the prevalence of industrial animal operations. Rather, they have gained market advantages and lowered their costs by exerting political and economic power. This is done in a number of ways.

One is by controlling livestock prices. Large vertically integrated meat companies now own the vast majority of the nation's slaughtering capacity. Family farmers have told me they could not get their livestock or poultry slaughtered unless they entered contracts with vertically integrated meat companies. Independent farmers are thus involuntarily moved out of the free market and forced to submit to contracts with agribusiness. These are lopsided deals in which farmers must accept whatever prices meat processors are willing to pay.

Agribusiness also lowers its expenses by deftly avoiding paying many of the true costs of doing business. Recall that in Iowa, the director of the Department of Natural Resources told Rick Dove and me that his skimpy budget made it impossible to enforce environmental laws and regulations against animal operations, even though it's mandated by federal law. Iowa is typical in this regard. For example, a study by the nonprofit Izaak Walton League documented that environmental enforcement is seriously deficient throughout much of Midwest farm territory. It concluded: "Every state program in the Upper Midwest is under-staffed and under-funded to provide the adequate oversight needed for livestock [operations]."

In my firsthand experience and research, I've come across no state or federal agency that is truly enforcing environmental laws against industrial animal operations. This is the result of decisions made by legislators and political appointees who are heavily under the influence of, and often come directly from, agribusiness. The ability to evade environmental enforcement translates to lowered costs for industrial operations. I've seen no specific account of this savings, but my guess is that it's substantial.

Additionally, agribusiness has long made use of its political power to obtain significant public subsidies. They are yet another way of foisting agribusiness' true operating costs onto the U.S. tax-payers. (I find it more than a little ironic that at the same time agri-business vociferously insists it employs the "most efficient" ways of producing food, it presses for ever greater public handouts). In 2005, Americans paid agriculture more than $21 billion in federal farm subsidies. Politicians are fond of portraying the subsidies as helping small family farmers. But from 2003 to 2005, 84 percent of federal farm subsidies went to the largest one-fifth of farms.

The federal farm subsidies are not paid directly to animal fa-cilities. Instead they are given to farmers growing designated com-modity crops, including corn and soybeans, the main sources of livestock feed. Except in the case of beef, feed is the single larg-est cost of raising animals. Specifically, feed costs now account for more than 60 percent of poultry and egg costs, almost 50 percent of hog production costs, and about 17 percent of beef cattle costs. Feed costs account for 46 percent of total costs in the dairy industry.

Dr. Timothy Wise, Deputy Director of the Global Develop-ment and Environment Institute at Tufts University, has closely examined the relationship between the U.S. animal sector and fed-eral grain subsidies. His research concluded that "[t]he corpora-tions that dominate the industrial livestock sector in the United States are among the chief beneficiaries of low U.S. feed prices." Dr. Wise has calculated that in recent years federal farm subsidies were giving agribusiness livestock operators "a discount of about 15 percent on their most important operating cost," and a total cost reduction of as much as 10 percent. Moreover, Dr. Wise believes that, overall, traditional farmers have been harmed by federal grain subsidies. Among other reasons, he notes, "when they feed their own livestock with a mixture of their own grains, they are in effect paying full cost for that feed, while corporate buyers are getting it below cost."

On top of such subsidies, agribusiness regularly seeks and is granted other forms of public assistance, including from state and local governments. For example, a Cargill-owned hog operation benefited when Colorado's Agricultural Development Authority secured $15 million in tax-exempt bonds to help finance its waste management facilities. In Kansas, Seaboard Farms (the nation's third largest hog producer) was provided $9.5 million in tax-exempt bonds for the same purpose. It hardly needs to be noted that no comparable public assistance is available to family farmers.

Moreover, if industrial operations' massive collections of animal waste were treated in the manner law and sanitation standards require for human waste (something that is arguably scientifically warranted), their costs would skyrocket. According to Dr. Weida's analysis, "the mid-range cost for waste treatment alone would be $173 per [hog] at the national average sewage disposal cost." He argues that to accurately compare efficiencies of animal operations, one must consider such waste disposal costs. "The fact that [industrial animal operations] try to avoid this cost by shifting the cost of their waste to the surrounding region makes no difference—the confined operation is still less efficient in an economic sense," Dr. Weida explains.

Stated another way, the environmental and public health problems caused by industrial animal operations' waste are real costs. It's just that instead of being borne by the companies, they are passed on to society. Dr. Weida and other economists refer to that as "externalizing costs."

The economies of traditional farms, on the other hand, are not based on externalizing their costs. Quite the contrary. For farms that keep their animals on pasture, manure plays a totally different role. Such farms do not need treatment facilities for their animal waste because they do not liquefy it and they produce much less of it. On well-run farms, animal waste does not cause pollution problems and is an integral part of an age-old "closed loop" sustainable farming system. Manure builds soil texture and replenishes nutrients, fertilizing crops grown to feed livestock, all of which makes manure an asset on a traditional farm, a valuable part of the farm's economic and environmental cycles. An old farm adage even advises: "You can measure a man's wealth by the size of his manure pile."

In conjunction with the myth of greater efficiency, I've often heard it asserted (again, with no proof) that industrial animal operations help rural economies. This also turns out to be a myth. The most comprehensive study on the question was carried out by University of Illinois researchers who scoured data from more than one thousand rural Illinois towns from a 17-year period. The study determined that industrial hog facilities actually harm rural economies, concluding: "[L]arge hog farms tend to hinder economic growth in rural communities." It also found that the more industrial hog production there was in a community, the lower the per capita retail spending.

In a related vein, several studies have demonstrated that industrial agriculture fails to support local economies with its purchasing. Researchers from the University of Minnesota Department of Agriculture and Applied Economics documented that the local farm expenditures of livestock operations actually fall sharply as their size increases. Industrial operations are more likely to source just about everything—feed ingredients, pharmaceuticals, protein sources, vitamins, and minerals—in bulk from distant dealers.

Industrialization also reduces jobs. The number of hog farmers, for instance, has dropped by 95 percent since the middle of the twentieth century, concurrent with spreading industrialization. The effect of industrialization on employment levels is such an important (and misunderstood) point that I deliberately invited an economist, Dr. John Ikerd, to address it at our North Carolina and Iowa Waterkeeper Hog Summits.

Dr. Ikerd, a retired agricultural economist from University of Missouri, has closely evaluated how industrialization has affected overall farm employment. He explains that, as in other sectors, industrialization in animal farming has meant a shift away from people (labor) and toward buildings and machines (capital). His research shows that "[l]arge-scale, specialized operations produce more hogs per person employed, and consequently, create fewer jobs per hog produced." Therefore, Ikerd explains, "large-scale, contract production employs far fewer people than would be employed to produce the same number of hogs" by independent farmers. More specifically, Dr. Ikerd has calculated that smaller scale hog farms employ almost three times as many people per animal as industrial hog facilities.

In fact, this is much of the true meaning behind the word *efficiency* in the animal raising context. When USDA states that industrial operations are more efficient than traditional farms, it really means that industrial facilities employ fewer people per animal raised and pay workers lower wages. Industrial operations have all but eliminated skilled animal husbandry, replacing it with crates, cages, pens, mechanized feeding and watering systems, and continual pharmaceutical dosing. Consider the following: USDA figures show that a pig at an industrial operation spends about eight *seconds* per day in the company of a human. In other words, the pigs are not really receiving any human attention at all. Thus, the "efficiency" of industrial operations is also externalized onto the backs of the animals.

For me, the way animals are treated is the most compelling reason to avoid buying the products of industrial animal operations. Animals matter to me and industrial operations are inherently inhumane. That is not an accusation that every person running or working in these facilities is an animal abuser. Many of them are caught up in a system over which they have little or no control. Yet that system was designed to maximize production while providing animals only the barest necessities for survival. Most animals can survive in industrial operations but it is literally impossible for them to have good lives.

The assembly lines of industrial systems function well for the mass production of inanimate objects. But they are complete failures at respecting the individuality, instincts, and needs of living creatures. Industrial facilities impose daily misery on animals and cause behaviors in many that are almost unknown in the wild. These so-called "vices"—behaviors like tail biting, head waving and chewing the air—are treated as a normal part of animal farming. But there's nothing normal about such aberrant behaviors. They are symptoms of a fundamentally broken system, one that cannot be fixed with tweaks.

In response, however, our publicly funded universities spend millions of research dollars studying just how to do such tinkering. They seek ways to maintain industrial systems while dealing with animal "vices" and the disease and pollution problems intrinsic to continual confinement systems. I have no interest in participating in that misguided cycle. I don't want to eat food coming out of industrial facilities and I don't want my money supporting them—either as a consumer or a taxpayer.

Of course, raising animals on smaller scale family-owned farms does not in itself guarantee that they will be treated humanely. But it certainly creates the opportunity and, in my experience, the likelihood. Most important, the traditional farms I know give animals much more hospitable physical environments: ones that just let pigs be pigs, chickens be chickens and cattle be cattle. They operate on a smaller scale and are run by people who actually own the animals and live in proximity to them, all of which means animals get more and better care. On these farms, as on our own ranch, animals are individuals, not numbers. With a little effort, consumers can find out which farms are raising animals the right way.

Another reason I avoid buying industrial animal products is

their widespread overuse of drugs, especially antibiotics, which is another way industrial operations lower their costs. We've seen that adding antibiotics to feed reduces total feed costs and helps stave off the disease outbreaks intrinsic to crowded, confined animal populations, thereby reducing economic losses from animal morbidity and mortality.

Yet society pays a high price for this antibiotics overuse because it contributes to the rise of disease-causing organisms that are resistant to antibiotics, diseases with which people come in contact through the animals and in the meat. The latest (and greatest) threat from drug resistant diseases is known as methicillin-resistant *Staphylococcus aureus* (MRSA). A 2007 article in the *Journal of the American Medical Association* estimated there were almost 100,000 human infections and nearly 19,000 deaths from MRSA in the United States in 2005.

European research has been documenting the rise of MRSA at confinement animal operations for several years. The UK Soil Association describes the situation in the following urgent terms: "Farm animal MRSA is spreading like wildfire on intensive farms in continental Europe. In the Netherlands it already affects 39 percent of pigs and almost 50 percent of pig farmers. In Dutch hospitals 25 percent of all MRSA cases are now caused by the farm-animal strain, and farmers are no longer permitted in general wards without prior screening." The Dutch Minister of Agriculture officially stated that "the high use of antibiotics in livestock farming is the most important factor in the development of antibiotic resistance, a consequence of which is the spread of resistant micro-organisms (including MRSA) in animal populations."

In 2007, research about North American animal operations brought the concerns closer to home. A study published in *Veterinary Microbiology* found MRSA prevalent in Canadian pig farms and pig farmers, pointing to animal agriculture as a source of the deadly bacteria. The study discovered MRSA at 45 percent of farms, in nearly one in four pigs and one in five pig farmers. Some nine million Canadian hogs are imported into the United States annually. Regarding the research, the nonprofit organization Keep Antibiotics Working stated: "The heavy use of antibiotics in industrialized livestock operations can select for resistant bacteria, such as MRSA."

I can't resist briefly noting the pork industry's response. An article in *National Hog Farmer,* titled "MRSA Scare Blown Out of Proportion," said: "Despite all the scare-mongering projected by Keep Antibiotics Working, there is no need to avoid pork consumption or worry that pigs could make you sick as a result of MRSA, according to National Pork Board staff veterinarians."

I suspect that as long as there are industrial animal confinement facilities, some form of antibiotic overuse along with rising levels of antibiotic resistant pathogens will persist. Certainly, the facilities will continue to stink and to pollute our air, water, and soils. I know they will never provide animals good lives. These are the predictable consequences of keeping huge numbers of animals crowded together, continuously confined, and they are some of the true costs of producing "cheap food" from animals.

THE ANTAGONISTIC EXTREMISTS

Over the years, I've heard just about every argument against reforming the food animal industry. Not surprisingly, the loudest squeals come from the industry itself, which tends to repeatedly trot out the same tired arguments opposing reform. The industry predictably begins by asserting that "there is absolutely nothing wrong with current practices." Then it launches into claiming, in essence, that industrial farming is "the only way to feed the world."

These are the people—usually representatives of the National Pork Producers Council, the National Egg Board, the National Chicken Council and even, tragically, our own Food and Drug Administration and U.S. Department of Agriculture—who vehemently defend every aspect of industrial animal operations and aggressively attack anyone who raises the slightest criticism. They are what I call "industrial extremists."

As industrial methods and their consequences are becoming more widely documented and publicized, that position has become untenable and will be even less credible in the future. Suffice it to say, I consider reform of food animal raising urgently needed.

However, the second point—that we need industrial agriculture to feed the world—does require a response. I've seen this "necessity" claim in scores of agribusiness press releases, speeches, brochures, and advertisements, used both as shield and sword. As a shield, it's employed to deflect just about any criticism of agribusiness; as a sword, it's used to attack anyone who advocates alternatives. As we've just seen, the suggestion that industrial operations are economically necessary holds no water. Related to this is the argument that industrialization is necessary to produce enough food for the human population.

A few years ago, I had a firsthand demonstration of agribusiness wielding the "necessity" sword. Bill and I were attending a food and agriculture conference in Washington, D.C. A Monsanto sponsored event, it was a large conference made up almost entirely of speakers from companies like Monsanto and Tyson or from USDA. Bill was part of the sole panel that dealt with traditional and organic farming, which was sandwiched between speeches by John Tyson and Ann Veneman, Secretary of Agriculture.

The moment that Bill's panel concluded several people ran to the audience microphones. The first to speak was a past president of the National Pork Producers' Council. "You folks don't seem to be at all troubled that what you're advocating for would mean starvation for millions of people!" he charged with all the moral indignation of Pat Robertson. I felt my eyes rolling. His tirade, decorated with plenty of platitudes and arm flailing, lasted for several minutes. Immediately following, two other agribusiness higher-ups each stepped to the mike to elaborate on why the humanitarian obligation to our fellow man mandates industrialized agriculture.

The spectacle was breathtaking. A group of hardworking people whose lives have been devoted to socially responsible food production were being vilified as callously indifferent to the world's hungry while agribusiness was held up as a shining moral beacon. The panelists rallied with an admirable defense. For his part, Bill pointed out that the same number of pigs are in the United States today as in 1915, a fact that clearly demonstrates the fallacy in claims that industrial confinement systems are needed to meet current demand.

But I was bowled over by the sting of the agribusiness attack. The clever strategy attempts to turn the moral equation upside down by making advocates for sustainable agriculture the ones who are morally suspect. For who could in good conscience argue *against* feeding millions of hungry people, even if the negative side effects were hefty? If industrial agriculture were actually necessary to avoid mass starvation, it would move beyond morally defensible to morally imperative.

However, the assertion that industrialized animal agriculture is "necessary" for feeding the world's population is a house of cards that cannot withstand even a breath of scrutiny. In particular, I have never seen a shred of evidence that meat from industrial operations is feeding the world's malnourished. And the suggestion that industrially produced meat, dairy products, and eggs can somehow solve world hunger is patently flawed. For starters, it has long been known that more resources are required to produce human food from animals than from plants. The argument has been well articulated, most assiduously by Frances Moore Lappe in her 1971 book *Diet for a Small Planet*, that as the global population increases, the world community must shift away from meat and dairy toward food from plants.

Going a step farther, a 2007 report by the nonprofit World Society for the Protection of Animals specifically links hunger to industrial animal production. It connected the rise of large, corporate controlled animal production to farmers the world over losing their land due to lower-priced competition. In Brazil, the globe's third largest poultry producer, industrial production caused twenty thousand farmers to leave the countryside in one recent year, according to the report. "We believe factory farming is one of the root causes of hunger and poverty in the world today," said a WSPA spokesperson. Even if producing more food were the answer to hunger, expanding industrial animal operations would be a step backward.

More important, the suggestion that more food production is necessary to feed the world is based on a faulty premise. That assumption is that world hunger is caused by insufficient global food production, which is easily proven untrue. As food and nutrition expert Marion Nestle notes in *Food Politics*, the global food system currently produces some 3,800 calories per day of food for every man, woman and child, which, she points out, is about twice what's necessary for adequate nourishment. In fact, Nestle argues that the "greatest unspoken secret" of the U.S. food system is "overabundance."

Global food production has actually outpaced population growth. Every year the world produces enough wheat, rice, and other grains to provide 4.3 pounds of food per person per day (including two and half pounds of grain, beans, and nuts, a pound of fruits and vegetables, and nearly a pound of meat, milk, and eggs). Moreover, in the last four decades, per capita food production has grown 16 percent faster than the world's population, meaning there is now more food per person available on the planet than ever before in history. Clearly, abundance is not the issue.

Dr. Nestle and other food experts argue that hunger results not from a dearth of food but from distribution failures and, even more important, from people lacking the resources to buy food. In other words, poverty. Martin McLaughlin, a former professor of food security, agrees. His book *World Food Security* repeatedly emphasizes that poverty is the root cause of hunger. "Hunger . . . is a political and social problem," he writes. "It is a problem of access to food supplies, of distribution, and of entitlement." The book *Fatal Harvest: The Tragedy of Industrial Agriculture*, makes the same point several times over, noting that even during the height of the 1980s famine, Ethiopia continued exporting substantial amounts of grain.

Moreover, there is credible evidence that industrial agriculture has made it harder for poor people to feed themselves. In the United States, industrialized agriculture and government policies have reduced the market access of smaller farmers. In developing nations, corporate farms have pushed up land values to the point that subsistence farmers can no longer afford to keep their lands, causing them to migrate en masse to urban areas. Large numbers of displaced farmers have joined the ranks of the world's landless poor who have little ability to buy food. More food production is no help to those who have no land to grow food and no money to buy it. The solution to world hunger is certainly not more of the same.

Agribusiness industrial extremists often make a related argument that it would be impossible to maintain current meat production levels without industrial operations. "There just isn't enough land to raise animals outdoors," the argument goes. The suggestion that we can raise all the farm animals we need on pastures is characterized as quaint, and supporters of traditional agriculture are characterized as nostalgic, or even Luddites. But again, the assertion that there's insufficient land for traditional farming is made without proof and there's no reason to believe it. Under the industrial animal system massive amounts of feed are produced by plowing, planting, and harvesting vast tracts of land then transporting the feed to the animal facilities. Waste is then transported away from facilities and applied to land. Obviously, those crop lands and waste disposal areas must be included in any land use comparisons.

Traditional farms with animals ranging outdoors actually use land more efficiently. Consider a pig farm like Paul Willis'. He raises his own feed crops and rotates his pig herd onto resting cropland. The pig manure helps restore the soils. Moreover, grazing pigs consume less feed and use it more efficiently. "With good clean ground, pasture hog gains may be produced on 15 to 20 percent less [soy and corn]," notes the authoritative treatise *Swine Management*. The book also states that "the better hog pastures replace one-half of the protein concentrate" and that often, "faster gains are made on pasture." All of this means that less land needs to be plowed and planted per pig at a farm like Paul's compared to an industrial operation. In short, farmers like Paul are making better use of land.

The claim by industrial extremists that agricultural industrialization is necessary to feed the world parallels its rhetoric justifying genetically modified organisms. A recent posting on the website of Monsanto (the world's top producer of genetically modified seeds) is typical: "GM Crops Can Contribute to Increased Food Production and Reduced Hunger."

Actually, there's scant evidence that genetically modified crops benefit the poor. In 2008, the nonprofit environmental organization Friends of the Earth reported that most genetically altered crops are not destined for hungry people in developing countries but are instead "used to feed animals, generate biofuels, and produce highly processed food products—mainly for consumption in rich countries."

For years, controversy has raged, especially in Europe, about injecting genetically modified crops into the natural ecosystem and our food system. Farmers point to studies showing that pollen can be carried many miles by wind and bees to contaminate non-modified crops. Consumers have voiced concerns over food safety. So much opposition exists in Europe that in 2002 a committee of the European Parliament adopted a strict traceability and labeling requirement for all foods from genetically modified crops. In response, the Bush Administration (and this was one of its major agricultural initiatives) sued the EU at the World Trade Organization court in 2003 to force Europe to accept our genetically altered commodities. The court case is ongoing.

Still more troubling is the genetic alteration and cloning of farm animals. Using taxpayer funding, scientists collaborating from several universities in the United States have been toiling away at doing things like taking genes from roundworms and splicing them with genes of pigs, then cloning these animals. The scientists are thus creating pigs whose genes have been tampered with on two levels. The purported purpose? Bacon with high omega-3 levels. I can't imagine a more unappealing or unnecessary food, especially in light of the fact that a person needing more omega-3s may be able to get more than enough by supplementing her daily diet with a couple of walnuts.

Moreover, according to the Humane Society of the United States, animal cloning technology is plagued with serious animal welfare problems. Cloning involves "high failure rates, birth deformities, disabilities, and the premature deaths of both surrogate mothers and their cloned offspring." For those reasons, the European Commission's European Group on Ethics in Science and New Technologies concluded in January 2008: "Considering the current level of suffering and health problems of surrogate dams and animal clones . . . [we do] not see convincing arguments to justify the production of food from clones and their offspring."

What's less certain but equally troubling are the long-term consequences to humans and the environment of genetically altering the plants and animals we raise for food, which are both unknown and unknowable. They are unknown because to date, no long-term epidemiological studies have been conducted. And they are unknowable because, as we've learned from past catastrophes—like the pesticide DDT and the drug DES—there are infinite and unimaginable ways that putting technological inventions into ecosystems and our bodies can affect life forms. Some of these unforeseen and unintended consequences might only become apparent over generations.

Nonetheless, in January 2008 the FDA declared that food from cloned animals and their offspring was safe. Even if the FDA is correct in this assessment (impossible to know at this juncture) the benefits of animal cloning are at best elusive. "Proponents of cloning technology say it could have a major impact on the livestock industry by providing meat and milk that is better and more consistent," according to an article about the FDA announcement. "When you buy a box of Cheerios in New York and one in Champaign, Illinois, you know they are going to be the same," said the president of an Illinois animal cloning company.

The most eloquent counterpoint I've seen to such vapid logic was in an essay by Verlyn Klinkenborg in the *New York Times*. He wrote:

> *I think the clearest way to understand the problem with cloning is to consider a broader question: Who benefits from it? Proponents will say that the consumer does, because we will get higher quality, more consistent foods from cloned animals. But the real beneficiaries are the nation's large meatpacking companies—the kind that would like it best if chickens grew in the shape of nuggets. Anyone who really cares about food—its different tastes, textures, and delights—is more interested in diversity than uniformity.*
>
> *As it happens, the same is true for anyone who cares about farmers and their animals. An agricultural system that favors cloned animals has no room for farmers who farm in different ways. Cloning, you will hear advocates say, is just another way of making cows. But every other way—even using embryo transplants and artificial insemination—allows nature to shuffle the genetic deck. A clone does not.*

Industrial extremists adamantly defend animal cloning in spite of widespread public discomfort with the practice. The credibility of everyone in agriculture gets tarnished every time such individuals and organizations make outlandish defenses of every agricultural practice or attack sincere critics and reformers. We in farming and ranching are poorly served by having industrial extremists serve as our purported spokespeople. It is my sincerest hope that more farmers and ranchers come to this realization and become the driving force for the needed overhaul of the industry.

On the polar opposite end of the spectrum from the industrial extremists are certain vegans. I've had many tolerant, compassionate vegan friends and colleagues over the years. Though it pains me to admit it, I must even put my ex-boyfriend into that group. I know much better than to paint all vegans with the same brush. But there is a decidedly radical element and its calling card is intolerance.

That fringe faction, which I call "vegan extremists," lashes out against anyone who accepts the idea of raising animals for food or fiber and is especially vicious when it comes to farmers and ranchers. Their most savage attacks are directed at the people they consider the ultimate hypocrites: fellow animal activists (such as Animal Welfare Institute and Humane Society) who support anything less than total "animal liberation" and those people engaged in animal farming and ranching who advocate for compassionate treatment of animals.

I had a close encounter with just such a vegan extremist in 2007 when I was invited to speak at an animal advocacy conference in Washington, D.C. At the talk, which was attended by nearly a thousand people, I presented pictures of our ranch and other farms and ranches I've visited. I spoke about the obligation humans have to take good care of every farm animal. As I do in all my speeches, I also stated my belief that industrial systems cannot give animals an acceptable quality of life. I shared the stage with pig farmer Paul Willis and turkey farmer Frank Reese, who each described and showed pictures of their own farms.

As soon as the last person on our panel finished, a woman charged to the audience microphone. "I don't know how you people can *live* with yourselves" she sputtered with an accusatory pointing finger. "You must be another species!" Her face reddened as she continued several minutes of vitriolic diatribe. She then turned around and shouted obscenities at the person waiting in line behind her who had suggested she allow others the chance to speak. Shortly after that, the woman was ushered out of the room by one of the event sponsors. I'm told she was shaking and in tears on her way out. The woman was literally enraged that farmers had even been allowed to speak at the conference.

Months later, it was brought to my attention that the same woman posted a photograph of me on her website. It was accompanied by a rambling article in which she decried the destruction of "the movement" by animal advocates seeking anything short of what she called "abolition." In mentioning me, she reiterated her disgust that I would be invited to an animal advocacy conference. She grouped me with a veal confinement operator who'd made a statement favoring the elimination of veal crates and categorized us both as hypocritical "butchers."

It's odd how offensive the vegan extremists find it when a person involved in animal farming expresses any support for humane animal treatment. Would they prefer that the people who raise animals for food show no concern at all for the creatures in their charge? Apparently so. It's more in keeping with the vegan extremist portrait of the world in stark black-and-white contrast, in which every person who accepts animals in the food system is morally flawed and everyone involved in raising animals for food is a charter member of the axis of evil.

All of us who care about animals are disadvantaged by the vegan extremists. Unfortunately for the rest of us, such extremists are often the public face of animal advocacy. Their radical claims and tactics get the most media attention, making many policy makers and members of the general public uncomfortable about any initiative intended to help farm animals. Agribusiness industrial extremists relish pointing to vegan extremists and claiming that they are the true backers of every effort for reform.

Most Americans are simply caught in the crossfire. Very few of us hold opinions represented by either the industrial extremists or the vegan extremists. Their polarized rhetoric about farm animals stymies reform and fails to offer constructive solutions.

As for my own view about the appropriateness of consuming animal flesh, it has been reaffirmed by my life on the ranch. Here we are surrounded by so much wildlife that I see animals eating other animals in the wild every day. I witness ospreys, hawks, bobcats, and coyotes hunting and feasting on their prey; I catch ravens slurping the contents of purloined eggs; and I see vultures and insects devouring carcasses of all shapes and sizes. Such scenes are daily reminders of nature's rhythms of living, dying, and gradually returning to the earth after death. Ashes to ashes and dust to dust. I accept the raising of animals for food and believe that good farming and ranching is in harmony with natural cycles.

I also believe that peoples' lives are deeply enriched by spending time in the company of animals and that the lives of animals are enhanced when they are in the care of people who genuinely look after their well-being. As Temple Grandin has said in her book, *Animals in Translation*, "Animals and people belong together." We have so much to learn from the animals and, if we're paying attention, they teach us every day.

For my part, I am tremendously thankful that there are people who share my own love for animals involved in farming and ranching. Since the vast majority of the world's population consumes meat and other animal products—and global consumption of animal products is *growing* not declining—it is essential that such people set an example of humane animal treatment for the world of agriculture.

Moreover, I believe that those farmers and ranchers who demonstrate a strong animal welfare ethic and serve as models to their peers may be the single best hope for improving the lives of farm animals in this generation and the next. It is probably recognition of this reality that led Matthew Scully, in his book *Dominion: The Power of Man, The Suffering of Animals, and the Call to Mercy*, to refer to supporting traditional family farms as "a decent compromise," although he personally follows a vegan diet.

Thankfully, neither Bill nor I have ever taken the extremist positions. If we had, I'm sure we would never have made it through that first dinner together. In our own household, there is virtually no tension over the eating of meat or lack thereof. To his credit, in our five years of marriage, Bill has never pressured me to give up my own vegetarianism. The closest he comes is when eating a piece of meat he's especially savoring, saying, "Mmm. I wish you could taste this." And I don't pressure him to refrain from eating meat.

As a practical matter, however, he is eating a lot less of it these days. I do about 99 percent of the cooking in our home (unless you count making coffee, in which case Bill's share might move from 1 percent to 3). And, (other than for our wonderful Great Dane, Claire de Lune), I don't cook meat. So, that leaves Bill the choice of cooking his own or going without. He's found himself surprisingly contended to do without meat most of the time. When I make a pasta marinara or a margarita pizza, he sometimes cooks up some chorizo and slices it on top. Usually, he just eats the meal as is, looks at me with those bedroom eyes and croons, "This is delicious, Porkchop."

CHAPTER TWELVE

Back on the Ranch

AND LOOKING AHEAD

These days, I'm well-settled into my life here on the ranch with Bill, Claire de Lune, and all our other animals. It is not a life I could have predicted for myself. But it is certainly one I cherish and am grateful for every day.

We welcome a steady stream of visitors to our ranch. Family and friends, to be sure, but also chefs, journalists, and fellow farmers and ranchers. All of them want to learn more about what we're doing here and how we're collaborating with like-minded people in agriculture throughout the country. Bill and I enjoy sharing our experiences and ideas.

Our visitors frequently ask me: "Wasn't it hard to move here from New York City?" I respond that it was not, an answer that seems to surprise most people. Although I do look back fondly to my time in New York, I don't miss it. I love the energy, culture, diversity, and amazing restaurants in and around Manhattan. But when I resided there I yearned to be far from asphalt and immersed in nature. Each traditional farm or ranch I visited was a refreshing oasis. Interestingly, my parents, who know me so well, have not been at all surprised that I adjusted so easily to ranch life. I think that to them, I was destined for it. I am far from where I grew up but very much at home.

I'm well-served by the knowledge I've gained about the many facets of animal farming. It helps me in daily ranch chores as well as in my public speaking and writing, all of which advocates for returning common sense and humanity to the way our nation raises farm animals. I've stayed in close contact with my old buddies Bobby Kennedy, Rick Dove, Diane and Marlene Halverson, Terry Spence, and the many other allies I've made in joining this cause. And my adventures in agriculture and food continue to take new and unexpected turns.

The most interesting twist has been Bill's departure from Niman Ranch, Inc. It unfolded as a classic scenario. Like other founders of alternative enterprises, those based on strong ethics diverging from the mainstream, he became embattled at the business as it grew. Bill was never willing to abandon or water down any of the principles with which he started. And he was dedicated to living up to what he believed were customers' expectations of how things should be done. The company steadily grew, yet it struggled financially. So the pressure from senior management to dumb it down intensified.

For several years, Bill was making his daily trek into the office with a certain dread. Every day might turn into a boxing match. Bill would often return home crumpled and bruised. He envied me for being the one to stay back and spend the day working on the ranch.

Then the final straw was laid. The board of directors arranged for much needed outside investment money and brought in managers from the mainstream meat industry. None of the new managers had ever lived a day on a real farm or ranch. Once they were in place, it became apparent that everything they were proposing was designed to make Niman Ranch look and function more like the rest of the meat industry.

Some of these changes related to the animals. One thing that particularly bothered Bill was an alteration of the cattle slaughter procedure. For well over a decade, the people who cared for the cattle everyday at the feedlot, usually Rob or Michelle Stokes, personally accompanied every load of cattle to the stunning area at the slaughterhouse. Bill and the Stokeses believed that this was the best way to ensure the animals were never mistreated and felt no anxiety or fear. The new management team considered this long-standing practice unnecessary and it was soon suspended.

Other changes related to the handling of the meat. Living alongside animals for decades, Bill never forgets that every piece of meat involves taking an animal's life, making meat terribly and uniquely precious. Bill feels that even if it costs more, meat should always be handled with the utmost care; he considers any waste of meat a sacrilege. That was part of the reason that Bill had long ago set up a company-owned meat cutting plant, staffed by highly skilled butchers. He also insisted the organization do its own distribution. These things allowed the company to keep total control over the meat from start to finish. Over Bill's strong objections, the new group eliminated the meat plant and outsourced the butchering and distribution.

In spite of his valiant efforts to stay upbeat, this was a dark time for Bill. Not that Niman Ranch was abandoning all of his core values. We both still consider it the best in the meat industry. But there were dozens of smaller decisions like these that, in the aggregate, were eroding some of the pride Bill had always felt about the company. It just wasn't the place he wanted to be anymore.

Deciding to leave a business he'd spent his entire adult life building was painful. Bill came to the decision slowly and only after much deliberation and soul searching. Financially, too, departing would be a strain. Contrary to popular misconception, Bill never made a pile of money with Niman Ranch. The truth is, he poured every drop of his sweat, blood, and tears into it for more than thirty years and got little in financial reward. Primarily, it was a labor of love. It was a mission. But things had reached a point where Bill's happiness depended on his leaving, and I wholeheartedly supported his decision.

From where I sit, it has already proven a wise choice. Bill's natural energetic, optimistic disposition has returned. He now spends much of each day doing what he loves best: working on the ranch. At the same time, he can't help being an entrepreneur and perpetually bubbles with ideas and plans. So it has not surprised me that he's also diligently toiling over his next national enterprise. It will raise the bar even higher for animal farming, setting a new gold standard for humane, environmentally friendly farming. Bill and I are working together on figuring out how to create such an organization, based on the guiding principle that animals should live as fully and naturally as possible at all stages of their lives. We're thinking of it as "the next generation of animal farming."

Part of our future plans involves remaking our own ranch. Like most other farms and ranches in the United States these days, ours had become specialized over the years. When he started farming, Bill raised goats for milk, chickens for meat and eggs, cattle and pigs for meat, and he tended trees for fruit and a substantial swath of land as a vegetable garden. His goal was self-sufficiency, including growing his own food. But as Bill's time was increasingly dedicated to running the business side of Niman Ranch, his own farm narrowed its functions. The orchard grew over with shoulder-high poison oak and brambles. The vegetable garden was abandoned. The ranch stopped raising chickens, pigs, and goats.

Bill and I aren't turning back the clock to the homesteading days, but we have begun making the ranch much more diversified, with multiple layers that connect to and support one another. We've already reclaimed the orchard, which now generates far more plums and apples than we could ever consume as fresh fruit. I've added pear and quince trees. By early spring, the orchard overflows with delicate, fragrant blossoms and buzzes with pollinators. A few months later, when the fruits begin coming into season, I start shifting into the preserving mode. For each of the past three years, I've spent several days cooking and canning plum jam, quince paste, and applesauce. The supply conveniently lasts about nine months, running out just around the time the first fruits are ripening.

Every year I learn a bit more about the wild foods here, too. For several months, starting in early spring, there are patches all over the ranch of a delicious naturally occurring green, miner's lettuce, which makes wonderful, earthy salads. Around the same time, whole fields of wild garlic burst into little snowdrop blooms. I've found it to be a tasty seasoning in salads, sauces, and omelets. By mid-summer, we get a good blackberry crop from bushes scattered around the ranch. And, slowly and cautiously, I'm learning where and how to forage here for wild mushrooms mid-winter. This year we had a huge porcini that we ate over the course of three days.

Our next step in improving the ranch has been a return to vigorously managed grazing. In the days when the Stokeses lived and worked on this ranch, they put a lot of effort into pasture improvement. They hand cut thistles, disseminated seeds of perennial grasses and clovers, and, using portable electric fences, ran a modern grazing system here not unlike the ones we saw in Wisconsin. But when the Stokeses moved on to other parts of the Niman Ranch enterprise, that level of daily grazing management came to a halt.

Since exiting the company, Bill has again turned his attention to our meadows. He's dusted off the solar electric fencing and added new fencing, which allows us to graze smaller sections of land and more frequently rotate the cattle from pasture to pasture. He acquired native grass seed from the National Park Service, which he hand sows at certain times of year in particular patches of our land. He pairs this with carefully timed mowing of undesirable and invasive species.

And once again, he has enlisted the help of Rob and Michelle Stokes. They left the company shortly after Bill's departure and began building their own herd of cattle and goats. Traditionally, cattle and goats (or sheep) have been grazed one after the other because of their complementary eating habits. Goats prefer very different plants than cattle, resulting in more evenly grazed pastures and, ultimately, more vegetation growing per acre. After months of brainstorming by phone and email, it was decided that the Stokeses and their goat herd would come down from their ranch in eastern Oregon to pass winter and spring at our ranch. The goats would benefit from the superior winter forage at our place and our pastures would be helped by the complementary grazing and the natural fertility their manure would add to the soil. As I write this, they've all been here for three months and it's working out very nicely.

We've also been tending to our gardens. Our beds of flowers, succulents, and herbs have been expanded and now surround the house on all sides. I try to keep the house full of cut flowers year-round. My hope is to get the vegetable garden going again one day soon, although for the time being, we buy our vegetables from our wonderful local farms.

We've recently decided to try pasture raising a small flock of traditional turkeys here. After reading a couple of books about it, we put in a phone call to Frank Reese. We wanted to get his advice and find out if we could buy some young breeding stock from him. Frank has generously shared his knowledge and agreed to help us get started with the right birds. This sort of openness is not necessarily typical of every business, but it's the kind of thing farmers traditionally do for one another. As I write this, Bill and I will be driving to Frank's farm in about a month to get our turkey lessons straight from the master and pick up our first set of newly hatched turkeys (poults).

Bill and I are also laying the groundwork for re-establishing a flock of laying hens. We've almost completed what one might generously call the "research and development phase." Fortunately, there are lots of creative solutions for keeping the birds safe in pasture and we're studying the alternatives. We've purchased books on pasture raising chickens, downloaded poultry housing plans from the Internet, and spoken with several farmers raising fowl in fields. Our set up will likely involve some sort of mobile coop. Whatever we end up with, it will allow the chickens to freely roam, graze, hunt for bugs, and dust bathe during the day then retire to a warm, safe coop with nests and perches at night. We plan to allow our favorite hens to raise their own young.

In addition to one day having our own flock of hens, my other as yet unfulfilled dream is to have my own milk cow. I'd been talking about this with Bill for years but it had always seemed unworkable. Modern dairy cows produce far, far more milk than one family could use and must be milked at least twice every day. And male calves would be a problem for us just as they are for everyone else. The resolution to the dilemma has become clear over time.

Most of the answer came to us in a visit to a family in Michigan, the Petersens, whose picturesque free-range pig farm is not far from my parents' home in Kalamazoo. They keep a reddish-brown cow with horns, a Devon, for milk. Recall that Devons are dual-purpose breeds, so the volume of milk they produce is not overwhelming and their calves can be raised as high quality beef cattle. The Petersons allow their Devon cow to keep her calf for as long as she wishes. With a suckling calf, her milk continues to flow yet the calf's nursing relieves the family of daily milkings and of excess milk. They take milk just when they need it. The Petersen's solution is so simple, sensible, and humane.

Bill and I now feel ready to find a heifer to rear as a milk cow. I've been looking into dual-purpose breeds that thrive on a diet of grass. My guess is we'll end up with a Milking Shorthorn, but perhaps a Devon or a Brown Swiss. We will seek one that comes from thrifty bloodlines with good feet and legs; and healthy, properly sized udders. Whichever one we choose, she will have a name, she will graze year-round and she will be allowed to rear every one of her own calves. I look forward to drinking the first glass of milk from a cow with such a life.

Scientific evidence has been piling up for years that raising animals on pasture generates food that is more healthful and nutritious. The Union of Concerned Scientists' *Greener Pastures* is just one of many such reports. These studies are important tools in building the case in favor of farming based on pasture versus industrialized animal production and I cite them in my own speaking and writing. But for me, such nutritional analyses are superfluous. I've been around enough animal farming to know the kind of care I want for the creatures providing me with my food.

The joy I feel in the presence of our animals seems to be shared by many of our neighbors and passers-by. At church or the post office people frequently stop me to say something about how much they enjoy seeing our cattle. "It's just so calming to be in their presence," I remember one woman saying to me. Friends and neighbors frequently ask me questions or tell me stories involving our animals. The community especially enjoys seeing the little ones appear every fall.

One of my chief complaints against industrial animal operations is that they deprive us all of the pleasure of seeing animals out on the land. Some may dismiss that as a trivial matter, but I believe it has profound significance. By taking animals off the land, erecting ugly metal buildings, and digging huge manure pits, industrial animal operations have dramatically diminished the beauty of our nation's rural landscapes.

Farm animals are hidden from our view. We can no longer watch them peacefully moving across green hillsides, jockeying for position at the feed trough, establishing dominance orders, skipping through fields, engaging in courtship rituals, or just joyfully playing. Witnessing such occurrences helped Americans develop an understanding of who these animals are and how they are connected to us. Losing access to those scenes has been a great loss to us all.

Like many of my neighbors in Bolinas, I've always considered it a treat to see critters on the land. As far back as I can remember I've scanned the landscape for domestic and wild animals on every drive through the countryside. When we were children, my sisters and I would point and shout out from the backseat: "Look—*pigs*!" "Over there—*horsies*!" Spotting farm animals was our major amusement on those long drives down to Ohio and back to visit our Uncle Jack.

I recently read a study that stated: "[P]asture-based farmers commonly say that their operations do not generate complaints, and neighbors are often enchanted by the sight of their grazing animals." I didn't really need empirical research to tell me that. Just thinking back to those childhood road trips, or pulling into the driveway of a grazing dairy like the Parises or a traditional hog farm like the Willises, I cannot stop myself from smiling.

Here at our ranch, cars frequently slow down to gaze at our cattle. More than a few stop to take photographs. Since Rob and Michelle have brought their several hundred goats down here, the gawking has reached extremes. People driving by just come to a dead stop to stare. Some even climb out of their cars and step right up to the fence to get a better look. We keep joking that we've created a traffic hazard. "What a beautiful sight!" a man shouted out to me just the other day.

It all reinforces what I've long believed about the connectedness between animals and people. People crave it and when they're deprived of it, they miss it. No matter how hard industrial agribusiness tries to push animals out of our views and our minds, they remain part of our collective consciousness. And that's a good thing.

A VISION FOR OUR FUTURE

Bill has always paid close attention to the source of his meat and has never been a meat glutton. Still, after all these years of eating vegetarian cooking every day he's more aware than ever that he doesn't need large quantities of meat to be satisfied. These days, he's happy with a much smaller portion size than he once was. He's also found that he doesn't need to eat meat every day, much less several times a day as he once did. He often explains to people that he prefers to eat "less and better meat." I believe that such an approach is part of the solution.

As we've seen, industrial animal operations are more heavily subsidized than traditional farms, they have better access to processing facilities and markets, and they externalize many of their true costs. For these and other reasons, animal products from traditional farms are often more expensive at the point of sale than those from industrial facilities. This leads many people to feel that they can't afford to make the switch.

In reality, most of us can afford it if we're willing to make a shift in attitude and emphasis. As Bill has discovered, meat can be moved painlessly from the center of the plate to the side by eating it less frequently and in smaller amounts. He is certainly getting what he requires nutritionally. Eating meat less frequently and in smaller portions makes it possible to budget the same amount while switching to meat from traditional farms.

Eating less meat is not such a radical notion. Really, it's quite a new idea that a big slab of meat should be consumed several times a day. Herbert Hoover promised "a chicken in every pot" (and "a car in every garage") in the elections of 1928 because it conveyed something desirable and out of the ordinary. It described a special Sunday dinner. Mutti once told me that when she and Vati were first married in the late 1950s she informed him they would not be eating meat every day. After evaluating their budget she had determined that meat was a luxury they could not afford on a daily basis. "That was normal in those days," Mutti explained. "People didn't expect to eat as much meat as they do now." Only in recent decades, with meat overly abundant and improbably inexpensive have Americans lost their view of it as something precious. We can only benefit by returning to that earlier view.

I also believe it behooves us all to think about spending a little more for the foods we get from animals. As the old saying goes, you get what you pay for. A close look at animal farming economics reveals that there's a reason cheap food is cheap. Americans spend a much smaller portion of their disposable incomes on food (averaging slightly less than 10 percent) than people in other parts of the world. We also spend much less now than we did before the industrialization of our food system. In 1933, Americans spent more than 25 percent of their incomes on food. The current lower percentage is often celebrated (especially by agribusiness and USDA) as evidence of the food system's "efficiency." The figures really illustrate the connection between the industrialization of our food supply and the lower food prices to which we've become accustomed.

In my experience, paying a little more for food isn't nearly as hard as it first seems. Mostly, it's a matter of a shift in thinking and budgeting. Between college and law school, I lived for a year in France, where people spend close to 14 percent of their incomes on food. Initially, I was stunned by the prices I encountered at the grocery store, cafes, coffee shops, and the *boulangerie.* Then I began to notice that much of what I was buying was a heck of a lot tastier than equivalent foods I'd been purchasing in the United States. In other words, I was paying more but I was also getting more for my money. I adapted my expectations and learned how to shop there. I probably also bought fewer articles of clothing and music cassettes. In spite of my meager student-teaching salary, I got along fine that year and ate very well.

I had a similar experience when first seeking out non-industrial dairy and eggs. There was some initial sticker shock. But the quality of what I was buying was so much better that I soon came to regard them as an entirely different products than commodity dairy and eggs and shifted my expectations about what it should cost. Plus, I felt so much better about what I was eating that the higher prices seemed well worth it.

In the years my life has focused on animal farming and the

foods it produces, I've had literally hundreds of conversations with people on the subject. Never once have I heard someone say they'd prefer their food to come from metal and concrete buildings rather than pastures. What has come through loudly and clearly is that the vast majority of Americans consider raising animals for food morally acceptable. Simultaneously, they believe all animals should be treated humanely.

Research bears out the prevalence of these attitudes. In a 2007 Michigan State University poll, more than 92 percent of people responded that they considered it important that animals raised for their food be treated humanely. Likewise, a 2005 survey conducted by Ohio State University researchers found that 81 percent of respondents agreed that the well-being of farm animals is just as important as the well-being of pets and 85 percent agreed that farm animals' quality of life is important. In a 2007 national survey carried out by Oklahoma State University, 76 percent of respondents said that animal welfare was more important to them than low meat prices.

I take heart in these findings, which confirm my own experience that the lives of farm animals do matter to the overwhelming majority of Americans. And these sorts of opinion polls tell me that as more people become aware of the way animals exist in industrial operations, there will be a growing push for reform and a growing consumer demand for foods from humanely raised animals.

There are actually plenty of promising signs for the future of traditional agriculture. The successes of farmers like the Willises, the Parises, the Klessigs, Frank Reese, Rob and Michelle Stokes, and Terry Spence are especially heartening. By working outside the established agribusiness infrastructure, they are laying the groundwork for a more sensible way our nation can produce food from animals without resorting to the shortcuts of drugs and industrial methods.

The Internet is a godsend for smaller farmers interested in delivering their products directly to consumers. When I learned that the Klessig family creamery was up and running, I was anxious to try their cheese and immediately set about placing an order. They have a lovely website that shows pictures of their farm and their animals, describes their philosophies, and presents their products. Every enterprising farmer in America can make use of this amazing tool. (I should add that the Klessig cheeses are simply delicious.)

I also find great hope in the growth of companies like Niman Ranch and Organic Valley. Both have high standards that must be met by every farm and ranch that supplies them. The standards are available for consumers to review. Such organizations are ideal for farmers who want to spend all of their time focused on farming and not worry about the distraction of marketing directly to consumers via the Internet or farmers markets. These networks allow farmers to remain independent and to get their products to mainstream markets. (I'm not at all troubled that such entities are not limited to providing "locally produced" foods. The idea that the United States ever had a "local economy" for meat and cheese is a myth).

For farmers willing to focus on selling directly to consumers, farmers markets are invaluable. Their explosive growth is yet another promising sign. In the late 1970s, there were around 350 farmers markets in the United States. Today, there are more than 4,400. That means that no matter where you live, there's a good chance there's one nearby.

I'm also pleased that the organic sector continues to expand by leaps and bounds. Although it's still less than 2 percent of the U.S. food market, organic farming is "one of the fastest growing sectors of U.S. agriculture, with sustained growth of approximately 20 percent per year for the last 15 years," according to a report by Colorado State University. "Sales were estimated at $1 billion in 1990 and reached $10 billion in 2003." I have my gripes with the organic regulations because they fail to require that animals truly be raised on pasture. Nonetheless, it's a sign that people are paying more attention to how their food is produced. And anything that moves agriculture away from its addiction to drugs and chemicals can only be regarded as a good thing.

There is also some reason for hope that animal operations may move away from grain and toward grass. Several factors, particularly the high cost of fossil fuels and the interest in alternative fuels based on grains (so-called "biofuels"), have dramatically pushed up the prices of corn and soy in the past couple of years. Higher grain and soy costs are shifting the economics of animal raising, making it less profitable at this moment in history to raise animals on a strictly feed based diet. Returning more animals to pasture or, in the case of cattle, keeping them on grass longer, is a likely result.

Some important public policy advances have also been made, although notably mostly through citizen initiatives rather than by actions of legislators or regulators. Shortly after quitting Waterkeeper, I drove my trusty, rusty old VW (which was still going strong) down to Tampa, Florida. There, I spent a week gathering signatures for a state ballot initiative to outlaw gestation crates for sows. It was backed by the Humane Society of the United States, and other animal organizations. The effort was an overwhelming success, making Florida the first state to outlaw sow crates. HSUS followed this with a similar campaign and victory in Arizona.

A coalition of organizations is currently pushing for a law that will essentially outlaw sow crates, veal crates, and battery cages for hens—here in California. Bill and I are actively supporting the California initiative in several ways. We believe the California proposal reflects the public's wishes for farm animals and expect it to pass by a wide margin.

In May 2008, animal activists and agriculture worked to together to get a bill through the Colorado legislature and signed by the governor that will ban crates for sows and veal calves. In June 2008, Congressman Christopher Shays (R-CT) introduced federal legislation to prevent cruelty to farm animals.

These are some of the many signs telling me that a new age of awareness and concern for animals in our food system is in the offing. While I fully accept the appropriateness of humans raising animals for food, I do not accept that humans have a right to treat animals cruelly in the process, least of all for the purpose of higher profits. I know in my heart that it is immoral to bring animals into the world and then keep them in a way that they know only suffering. Yet that is the lot of the billions of animals being raised every year by industrial operations. If we wish to hold ourselves up to the world as a moral people, it cannot continue.

I am fully aware that not every concern I have about animal farming is likely to be addressed in the foreseeable future. It's unlikely, for example, that chickens and turkeys will hatch their eggs and raise their chicks and poults, or dairy cows will ever rear their calves to maturity. But meaningful steps can be taken in that direction. I know poultry farms that do some natural hatching and pasture dairy farmers that keep calves with their mothers for the first thirty days or longer.

What is certain is that the industrial system simply asks far too much of our farm animals. It gives them nothing, or perhaps worse than nothing, in return for taking their lives. This needs to change. Animals raised for our food should be provided a life that is worth living. We owe them at least that. And it doesn't take a team of agribusiness animal scientists to tell us what that looks like. It's a matter of common sense. We all know it in our guts and in our hearts.

We can also look to some of the world's great religions for inspiration. They teach that caring for God's creation includes good stewardship of *all* animals.

Buddhism is practiced in many different forms but is universally based on respect for all forms of life. Many Buddhist rituals conclude with the following words: "May all beings be free from enmity; may all beings be free from injury; may all beings be free from suffering; may all beings be happy."

Mohammed taught that there is a reward in doing good to every living thing. Modern Islamic scholars say that the prophet "reflected the Qur'anic teachings about the treatment of the natural world in his daily life. He encouraged the planting of trees, banned destroying vegetation even during war, loved animals and displayed great kindness toward them, and encouraged other Muslims to do likewise." A *hadith* of Mohammed says: "If without reason anyone kills a sparrow, or a creature lesser than that even, the living creature will put his complaint to God on the Day of Judgment, saying: 'So-and-so killed me for no purpose.'"

Jewish laws repeatedly command good treatment of animals. There are provisions forbidding eating before feeding one's beasts. There is even a provision that if one sees horses drawing a cart up a steep hill that they cannot manage, one has a religious obligation to help push the cart lest the owner beat the animals. Treating animals humanely is considered part of the general obligation to be merciful and compassionate.

Judeo-Christian texts are infused with lines urging respect toward all living creatures and attributing wisdom to them. The Book of Job advises: "[A]sk the beasts, and they will teach you; the birds of the sky, they will tell you; . . . the fishes of the sea, they will inform you. . . . In His hand is the life of every creature and the breath of all mankind." The Psalms are especially rich in references to animals' worth and wisdom. One of the Psalms, speaking on behalf of God, instructs: "For every animal of the forest is mine . . . I know every bird in the mountains, and the creatures of the field are mine."

Perhaps the greatest Christian example for compassionate treatment of animals was set by St. Francis of Assisi, who spent a lifetime communing with and helping all sorts of creatures. He included the following words in a prayer: "Be praised my Lord with all your creatures."

My favorite New Testament reading is in Luke, where Jesus teaches: "Are not five sparrows sold for two pennies? Yet not one of them is forgotten by God." What could more poignantly illustrate that the economic value we humans place on an animal has no bearing on its true worth?

Numerous Christian denominations have formally recognized a strong human responsibility toward all creation, even specifically endorsing traditional farming as the best way for humans to achieve this obligation. For example, an official statement of the United Methodist Church says: "We support a sustainable agricultural system that will maintain and support the natural fertility of agricultural soil, promote the diversity of flora and fauna, and adapt to regional conditions and structures—a system where agricultural animals are treated humanely and where their living conditions are as close to natural systems as possible."

I enjoy envisioning our world revising its attitude toward animal farming and truly heeding the teachings of these great religions. What if we just decided, as a people, as a nation, to ask a little less of the animals whose lives we take for food? What if we decided we wanted each of them to have worthwhile lives in which they are not only free from suffering but can experience daily joy? What if we decided that every animal deserves to feel soil under its feet, fresh breezes on its face, and sunshine on its back? What if we decided that we don't need a hen to give us quite so many eggs or a cow to give us quite so much milk? What if we decided that every animal should be bred to have sound limbs and body without oversized parts? And what if we decided to embrace the inherent diversity of animals and the foods they, as living creatures, provide us rather than trying to turn them into mono-genetic identically formed mass produced widgets? I, for one, believe we'd end up living in a far better world.

I suspect that few people reading these words will start picketing their grocery stores or running for Congress for the purpose of changing how farm animals are raised. As passionately as I feel, I'm not doing those things myself. But I have committed myself, for life, to the cause. I choose my foods carefully, continually seeking ways to do better, and I take every opportunity to educate people on what's really happening. I'm hopeful, too, that Bill and I are setting an example here on our ranch for others in agriculture.

To my fellow humans, I make the following plea: Do not thoughtlessly eat foods from animals. Know the source. Question the methods. There is great power in posing the following simple question to grocery stores, restaurants and farmers: "How was this raised?" Then shift your buying toward those meats, fish, eggs, and dairy products that come from animals raised in a way that you like. For far too long, stores, restaurants, and farmers have been able to tell themselves that no one was concerned about the treatment of farm animals. Just knowing that people care has the potential to spark massive change. It's voting with your fork.

"Eating is a moral act," says my good friend Brother David Andrews. We must be the change we seek in the food system, he urges. Long ago, I lost my taste for eating sanctimoniously and self-righteously. But I do believe that we can revolutionize the food system with righteous eating. If you're buying your pork from a farmer raising pigs on pasture, you're well on your way.

ACKNOWLEDGMENTS

First and foremost, I have to thank my loving husband, Bill, who has not only supported me in every possible way through the years of putting this book together, he also read countless drafts and offered innumerable useful comments. In so many ways, this book could not and would not have happened without him.

Secondly, I want to thank the people who provided useful comments and ideas throughout the process of writing this book. In particular, I am indebted to my agent Jennifer Unter who gave me invaluable advice from start to finish. My editor, Anne Cole, made many important improvements to the manuscript and was always responsive and reliable. I am also deeply grateful to Diane and Marlene Halverson, Ron Willis, Gabriele Hahn ("Mutti"), and my siblings, Christine Hahn, Robert Hahn, and Sigrid Hahn. All provided useful suggestions to the text and unstinting encouragement along the way.

Many other people, too numerous to name, have answered questions for me or helped me in my research. I greatly appreciate the help each of you gave me. In particular, I want to thank Paul Shapiro and Rick Dove.

I want to thank Bobby Kennedy for writing the foreword, for being an excellent teacher and mentor, and for having the foresight to put me on the path of this incredible journey.

I also greatly appreciate the many farmers who helped me along the way by answering my many questions and spending time with me on their farms and ranches. I am especially indebted to Paul and Phyllis Willis, Trish and Bert Paris, Karl and Robert Klessig, Elise Klessig Heimerl and Jerry Heimerl, Frank Reese, Terry Spence, Rob and Michelle Stokes, and Chuck Stokes.

And finally, I must acknowledge the help of Claire de Lune, the most faithful and loving dog a person could hope for. She kept me company during every stage of every draft and never lost faith in me for a moment.

NOTES

Chapter 1: My Crash Course in Modern Meat

3 *Premium Standard Farms*: was acquired by Smithfield Foods, Inc., in May 2007.

5 *Charlie explained that he'd been representing Terry and several neighbors against PSF:* As this book went to press, the case was still in litigation.

10 *pollution from industrial animal operations [is] one of the United States' most serious water pollution problems:* The U.S. Environmental Protection Agency would later release the following statement: "Wastes from factory farms are among the greatest threats to our nation's waters and drinking water supplies." "USA: Environmental Agency to toughen up on Manure," March 27, 2002, available at: http://www.just-food.com/article.aspx?ID=72059.

11–12 *For most of the twentieth century, the state's many farmers raised . . . a steady number of hogs:* Specifically, farmers raised 1.4 million hogs in North Carolina in 1900 and 1.9 million in 1981. USDA data, accessed May 2007.

16 *"Hydrogen sulfide (H2S) is a very poisonous gas . . .":* Field, B., "Beware of On-Farm Manure Storage Hazards," *Rural Health & Safety Guide*, Purdue University Cooperative Extension, from www.ces.purdue.edu, accessed June 2003.

16 *one hundred . . . studies linking . . . industrial animal operations to various ailments:* Personal library of author. Friends and colleagues have helped me collect these studies from all over the world. I am particularly indebted to Karen Hudson for her contribution.

16 *Workers are the most frequent human casualties of confinement fumes:* "As many as 30 percent of people working in swine confinement buildings suffer one or more chronic respiratory illness." Donham, K., "The Impact of Industrial Swine Production on Human Health," *Pigs, Profits, and Rural Communities*, State University of New York Press, p. 74 (1998). See also: "There is now an extensive literature documenting acute and chronic respiratory diseases and dysfunction among workers, especially swine and poultry workers, from exposures to complex mixtures of particulates, gases, and vapors within [concentrated animal feeding operations]. Common complaints among workers include sinusitis, chronic bronchitis, inflamed mucous membranes of the nose, irritation of the nose and throat, headaches, muscle aches and pains." *Iowa Concentrated Animal Feeding Operations Air Quality Study*, Iowa State University and the University of Iowa Study Group, p. 5 (February 2002).

16 *Chronic respiratory ailments are . . . hog operations' single biggest animal health problem:* "The structure of swine production has changed substantially in most swine producing areas over several years; large groups of animals are housed under intensive conditions, often in regions with an extremely dense pig population. High stocking density in a closed environment facilitates transmission of airborne pathogens within the herd and between herds as well. Consequently, respiratory disorders and systemic airborne diseases are today regarded as the most serious disease problems in modern swine production." Sorensen, V., et al., "Diseases of the Respiratory System," *Diseases of Swine*, Blackwell Publishing, 9th Edition, p. 149 (2006).

16 *Asphyxiations in hog confinements are commonplace:* "In a fully loaded, completely enclosed [confinement swine] unit, the CO2 level can rapidly rise and, along with depletion of oxygen, create an asphyxiate atmosphere in as little as six hours. The most obvious situation is a power failure, with ventilation completely down; however, there are other situations in which there may not be a power failure and there is still ineffectual ventilation." Donham, K., "The Concentration of Swine Production: Effects on Swine Health, Productivity, Human Health, and the Environment," University of Iowa, *Toxicology*, vol. 16, no. 3, p. 565 (November 2000). Over the years, I've collected articles on such incidents, including the following from *Pioneer Press*, June 2, 2004: "More than 1,000 head of 240-pound hogs died during the storm southeast of Lismore. . . . The hogs, which died from suffocation, were housed in barns that did not have natural ventilation."; and www.theiowachannel.com , July 5, 2004: "An electricity failure that reduced circulation and raised the temperature in a hog confinement is to blame for the death of 138 hogs at a farm in southeast Iowa. . . . This is the second incident involving dead hogs in recent days in southeast Iowa. About 600 hogs died of suffocation at a farm in Rome over the weekend."

16 *80 percent of the nitrogen in manure lagoons ends up in the air:* Jackson, L., "Large Scale Swine Production and Water Quality," University of Northern Iowa, in *Pigs Profits and Rural Communities* (1998).

16 *Ammonia is continuously emitted from . . . hog buildings where liquified manure is spread:* January 2001 Report of the U.S. Environmental Protection Agency on Concentrated Animal Feeding Operations, p. 13-18.

17 *North Carolina's hog operations alone were putting between 55,000 and 72,000 tons of ammonia nitrogen into the air every year:* "NH3 [ammonia] emissions in a six-county (Bladen, Duplin, Greene, Lenoir, Sampson, Wayne) area of North Carolina that maintains the state's densest and largest population of hogs increased significantly during the same period that the hog operations increased. Mean NH3 emissions from hog operations increased 316 percent between 1982–1989 and 1990–1997, 84 percent of the growth from all sources (i.e., hogs, fertilizer, cattle, turkeys, broilers, chickens) can be attributed to the increase in number of hogs." Aneja, V., et al., "Atmospheric Nitrogen Compounds II: Emissions, Transport, Transformation, Deposition and Assessment," *Atmospheric Environment*, vol. 35, pp. 1903–1911 (2001).

17 *60 percent of the nitrogen entering North Carolina coastal waters, including the state's fragile estuaries, comes from the atmosphere:* Aneja, V., et al., "Measurement and Analysis of Atmospheric Ammonia Emissions from Anaerobic Lagoons," *Atmospheric Environment*, vol. 35, at 1950 (2001).

17 *The air to water pollution pathway has been connected to algae blooms:* Aneja, *Atmospheric Environment*, vol. 35, pp. 1905–6.

17 *880 million pounds of nitrogen from liquefied manure spread on land, ends up in surface waters:* EPA CAFO Report, January 2001, p. 12–16.

17 *Hog operations also emit 70,000 tons of hydrogen sulfide gas, 296,000 tons of methane, and 127,000 tons of carbon dioxide every year:* EPA CAFO Report, January 2001, p. 13–19.

18 *Dumping waste into the atmosphere, it seems, has long been part of agribusiness' plans:* An example of a document that illustrates this industry "air dumping" strategy is the "Swine Manure Management Systems in Missouri," from the University of Missouri Agricultural Engineering Extension, by C. Fulhage & D. Pfost, accessed June 2004 at www.muextension.missouri.edu . It notes, "As shown in Table 1 for lagoons, 70 to 85 percent of the estimated input nitrogen can be lost in a lagoon, but this can be an advantage if limited land is available for manure disposal."

19 *almost all confinement animal operations continuously administer antibiotics in feed or water:* It is impossible to know with certainty how many operations add drugs to their feed because the government does not require reporting of such practices; and the industry (not surprisingly) has refused to disclose this figure to the public. However, a comprehensive report by a nonprofit organization, the Union of Concerned Scientists, estimated in 2001 that the majority of operations add antibiotics to hogs' daily rations. Mellon,

M., *Hogging It: Estimates of Antimicrobial Abuse in Livestock*, report of Union of Concerned Scientists, p. 40, January 2001.

19 *"The intentional mixing of water and animal wastes ...":* Halverson, M., "What's Wrong with Liquefied Manure?" *The Price We Pay for Corporate Hogs*, a report of the Institute for Agriculture and Trade Policy, sec. 3, 2000.

19 *"Unlike the composting or heating . . .":* Halverson, M., *The Price We Pay for Corporate Hogs*, sec. 3.

19 *Lagoons, then, are hazardous both in the quantity of manure stored and in the quality in which it's stored: Spills and Kills:* See *Manure Pollution and America's Livestock Feedlots*, a report by the Clean Water Network, Izaak Walton League, and Natural Resources Defense Council, August 2000, at p. 63, which notes the health and environmental benefits of raising pigs on straw-bedding and pasture. The listed benefits of using deep-straw bedding include: "Maintains manure in solid form allowing for safer storage and composting; Composting stabilizes the nutrients and reduces the volume of waste; Reduces risk of runoff when manure is applied to land since the waste in a solid form; and Reduces odor problems associated with liquid manure systems." Likewise, the report lists the following benefits of raising pigs on pasture: "Manure management and storage requirements are minimized due to spreading of manure evenly over pastureland as animals graze; Reduces odor problems; and Reduces soil erosion and water contamination."

20 *"The Neuse feeds into estuaries ...":* The Neuse River estuaries are where the river meets the Atlantic Ocean and are brackish, with tides going in and out each day.

21 *56 percent of Neuse River nutrient contamination is from agriculture: Neuse River Nutrient Sensitive Waters (NSW) Management Strategy,* North Carolina Environmental Management Commission (1997).

21 *Dr. Burkholder had also documented a 500 percent increase in ammonia in the Neuse:* Personal communication of author on June 11, 2004, with Dr. Michael A. Mallin, University of North Carolina.

22 And the Waters Turned to Blood: Barker, R., Simon & Schuster (1997).

23 *manure lagoons continually leach pollutants:* "In a set of 11 North Carolina swine lagoons, Huffman and Westerman (1995) found average inorganic (ammonium and nitrate) N concentrations of 143 mg/L in nearby groundwater, and found that through leakage the lagoons exported on average 4.7 kg N/day to groundwater." Mallin, M. & Cahoon, L., "Industrialized Animal Production—A Major Source of Nutrient and Microbial Pollution to Aquatic Ecosystems," University of North Carolina at Wilmington, *Population and Environment*, vol. 24, no. 5, p. 377 (May 2003). See also, a study that reviewed lagoon leakage showing that the rates varied depending on lagoon construction and topography, but all leaked. Ham, J. & DeSutter, T., "Toward Site-Specific Design Standards for Animal-Waste Lagoons: Protecting Ground Water Quality," *Journal of Environmental Quality*, vol. 29, no. 6, November-December 2000.

24 *leakage ends up contaminating someone's well:* "As of 1998, close to 1,600 wells located near factory farms in North Carolina were tested for nitrate contamination. Thirty-four percent of the wells showed nitrate contamination; ten percent of the wells had a nitrate level that exceeded the drinking water standard. The state's Department of Health and Human Services stated that the cause of the contamination was leaking hog lagoons and hog wastewater sprayfields." Marks, R., *Cesspools of Shame: How Factory Farm Lagoons and Sprayfields Threaten the Environment and Public Health*, report of the Natural Resources Defense Council, p. 37 (July 2001).

24 *Every year, throughout the country, many lagoons burst or flood over:* "Spills and dumping of manure and other waste products occurred over one thousand times at [concentrated animal operations] in the ten surveyed states between 1995 and 1998. Over 13 million fish were killed in more than two hundred of the spills." *Spills & Kills: Manure Pollution and America's Livestock Feedlots*, a report by the Clean Water Network, the Izaak Walton League of America, and the Natural Resources Defense Council, p. 1 (August 2000).

24 *According to university researchers who studied the spill:* Burkholder, J., Mallin, M., et al.. "Impacts to a Coastal River and Estuary from Rupture of a Large Swine Waste Holding Lagoon," *Journal of Environmental Quality*, vol. 26, no. 6 (November-December 1997).

25 *confinement hog operators funded a $30,000 study:* The Frontline Farmers report, entitled "Investigations of Water Quality for Selected North Carolina Basins," was authored by Dwayne R. Edwards, an agricultural engineer at University of Kentucky. To prepare the report, Edwards apparently merely reviewed data provided to him by Frontline Farmers. This was followed up by an analysis by Dennis Avery and his son and originally published by the Hudson Institute. Currently, it is posted by the Heartland Institute, (which spends much of its time denying the existence of global warming). The report is: Avery, D., & Avery, A., "Hog Farms No Threat to North Carolina Water Quality," March 2004, available at: http://www.heartland.org/Article.cfm?artId=14556 .

25 *In response, the state DNR pointed out that:* Barnes, "Hog Farmers: Rivers Unhurt," *Fayetteville Online,* June 9, 2004.

25 *Dr. Shower's research traces industrial animal pollution using a nitrogen isotope found only in animal waste:* Barnes, G., "Hog Country: Black River a Symbol for Both Sides," *The Fayetteville Observer*, December 17, 2003.

26 *Mallin and his team found a 265 percent increase in ammonia in the Black River:* Personal communication to author on June 11, 2004, by Dr. Michael A. Mallin, University of North Carolina.

28 *the largest hog complexes can actually generate more animal waste every day than the human feces produced in New York City or Los Angeles:* Dr. Mark Sobsey, University of North Carolina at Chapel Hill, available at: http://www.neuseriver.org/images/HOG_FACTS_final_with_clip_art.doc.

Chapter 3: What Came First, the Chicken . . . and the Egg

40 *The Boston Poultry Show of 1849:* Smith, P. & Daniel, C., *The Chicken Book*, University of Georgia Press, pp. 207, 218 (2000).

40 *major universities were offering classes in poultry husbandry:* Sawyer, G., *The Agribusiness Poultry Industry: A History of Its Development*, Exposition Press, p. 19 (1971) and Smith & Daniel, p. 246.

40 *between 1870 and 1926, more than 350 different periodical publications were introduced:* Sawyer, p. 21.

40 *chickens remained virtually omnipresent occupants of rural and urban America well into the twentieth century:* Sawyer, pp. 23–4; Smith & Daniel, p. 241.

40 *in urban areas there was one chicken for every two people:* Smith & Daniel, p. 233.

40 *the 1910 survey noted that 88 percent of farms kept chickens:* From: *Poultry Production in the United States*, report of the Economic Research Service/USDA, available at: http://151.121.68.30/publications/aib748/aib748b.pdf .

40 *with flocks averaging around 80 birds:* Smith & Daniel, p. 232.

40 *"the great era of the* farm chicken": Sawyer, p. 35.

40 *"the tranquil period in American life . . .":* Sawyer, p. 35.

41 *Big Dutchman feeders:* Sawyer, p. 164-5.

42 *"Fowls should not be kept . . .":* (emphasis in original) Wright, L., *The Practical Poultry Keeper*, Orange Judd Company, p. 1 (1867).

42 *"the hens will often drop their eggs . . ."* Wright, p. 31.

42 *Two calculations measured success in the newly emerging commercial poultry business:* Sawyer, pp. 97–98, 216.

43 *Around 1880, replacing mother hens with incubating machines began in earnest:* Sawyer, pp. 26–7.

43 *The incubator set poultry farming "on an unerring industrial course.":* Sawyer, p. 27.

43 *Humans first domesticated them:* This is based on the recent DNA analysis of a team of Japanese researchers, published by the National Academy of Sciences in 2006. However, when and where the chicken was first domesticated has been a matter of some debate. Many earlier sources claim that the chicken was originally domesticated in China or India some 4,000 to 5,000 years ago.

44 *Consider Plutarch, who often wrote admiringly about chickens:* Smith & Daniel, p.160.

44 *And listen to the raptures of Oppian, a poet of ancient Greece:* quoted by Smith & Daniel, p. 159.

45 *A single incubating machine could hatch 52,000 chicks at once:* Charles, T. & Stuart, H., *Commercial Poultry Farming*, Interstate Printing Company, p. 24 (1936).

45 *drowning two million male day-old chicks every year:* Smith & Daniel, p. 260.

46 *typical death losses quadrupled, jumping from 5 percent to 20 percent:* Smith & Daniel, p. 258.

46 *"With commercialization and greater intensification . . .":* Charles & Stuart, p. v.

46 *In everything he did, Salsbury encouraged liberal use of medications:* Sawyer, pp. 63–5. His company, eventually called Salsbury Laboratories, later became a part of the drug company American Home Products.

47 *Arsenic, of course, is a poison:* Arsenic is still widely (and legally) added to poultry feed in the U.S. The practice is illegal in the European Union because of concerns over environmental contamination and data connecting it with human cancers.

47 *"Food will win the war . . .":* Matusow, A., *Farm Policies and Politics in the Truman Years,* Harvard University Press, p. 3, (1967).

48 *"Those who could contribute to the war effort . . .":* Sawyer, p. 77.

48 *"[I]n every county across America . . .":* Sawyer, p. 77.

48 *Clementine Paddleford provides an illuminating report:* cited by Sawyer at p. 80.

49 *And farmers, who were famously protective of their independence:* Paarlberg, D., *American Farm Policy: A case study of centralized decision making,* John Wiley & Sons, Inc., chapters 1–3 (1964).

49 *Between 1929 and 1932, prices received by farmers fell by 56 percent:* Paarlberg, p. 17.

49 *maximum production was encouraged:* Paarlberg, p. 27.

49 *Truman administration maintained price supports:* Matusow, A., 1967.

50 *"Jewel controlled every phase of his operation . . ."* Sawyer, p. 91.

50 *extension agents actively supported entrepreneurs' efforts at vertical integration:* Horowitz, R., speech to Yale University Chicken Conference, New Haven, Connecticut, May 18, 2002.

50 *"laid the foundation for an agribusiness concept . . .":* Sawyer, p. 143.

51 *In each state, the industry's growth mimicked the pattern established by Jewell:* Sawyer, p. 138.

51 *the hen's egg production is related to light's effect on the pituitary gland:* Smith & Daniel, p. 264.

52 *"[W]hat they did was to organize the hens in a production line . . .":* Sawyer, p. 216.

52 *"Now the hen was to have a much bleaker life . . .":* Smith & Daniel, p. 268.

52 *Yolks lost their omega-3s:* Robinson, J., "The Benefits of Pastured Poultry," *Why Grassfed is Best,* p. 31 (2000).

53 *egg producers began adding red dye to hen feed to make yolks yellow:* Mallet, G., *Last Chance to Eat: The Fate of Taste in a Fast Food World,* W.W. Norton & Co., p. 285 (2004).

53 *"Industry representatives and academics agree . . .":* Pew Commission on Industrial Farm Animal Production, March 2008.

53 *The egg production cycle lasts about one year:* Government of Saskatchewan

website, at: http://www.agriculture.gov.sk.ca/Introduction_Poultry_Production_Saskatchewan.

53 *"Spent hens" are frequently vacuumed up into trucks and dumped into a rotating blade chopper:* Duncan, I., "Animal Welfare Issues in the Poultry Industry," *Journal of Applied Animal Welfare Science,* vol. 4, no. 3 (2001).

53 *A Georgia outfit raised eyebrows in the early 1950s:* Sawyer, p. 219.

54 *"[chicken] production is the prototype..."* words of Earl F. Crouse, as quoted by Sawyer, p. 172.

54 *The Truman administration had reasons, both practical and political, to support farm policy resulting in surplus grain:* Matusow, A., pp. 20–37.

54 *By the late 1950s, these technologies collectively increased the per-acre yield:* Paarlberg, D., p. 38.

55 *corn and wheat are heavily fertilized crops:* In 1999 corn accounted for 63 percent of all fertilizer used in the United States and wheat accounted for 24 percent. U.S. Environmental Protection Agency, Emissions Inventory, Oct. 27, 1999.

55 *farmers made little use of manufactured nitrogen fertilizer for some hundred years:* Beaton, J., "History of Fertilizer," Efficient Fertilizer Use Manual, www.mossaicco.com, p. 7, accessed May 2007.

55 *By 1945, artificial fertilizer use on the farm had almost doubled from just six years earlier:* Paarlberg, D., p. 36.

55 *When overproduction resulted in plummeting chicken prices in 1961:* Sawyer, p. 204–5.

55 *"Nearly 95 percent of commercial broilers..."* Western, J., *Wall Street Journal,* August 12, 1963, cited by Sawyer at 206.

56 *By the 1970s, farmers who were not under contract:* Lotterman, E., "Why No One Mourns the Loss of the Family Chicken Farms," www.fedgazette.com , April 1998; and Sawyer p. 219.

56 *"with the growing economic power imbalance in contracting..."* Taylor, C.R., "Restoring Economic Health to Contract Poultry Production," Agricultural & Resource Policy Forum, Auburn University, p. 6, 2002, available at: www.auburn.edu/~taylocr/topics/poultry/poultryproduction.html.

56 *"The silent and slow killer..."* Taylor, p. 3.

56 *"profitability has decreased to the point..."* Taylor, p. 2.

56 *"By the mid-1950s... the old patterns of selling..."* Sawyer, p. 170.

57 *by 1992, only 6 percent of farms had any poultry at all:* Poultry Production in the United States, report of the Economic Research Service / USDA, available at: http://151.121.68.30/publications/aib748/aib748b.pdf .

57 *"Although the exodus from agriculture in the past decade..."* An Adaptive Approach to Agriculture, from Economic Report of the President, 1967, pp. 236–7.

58 *Americans' consumption patterns closely track poultry farming's industrialization:* Data from the U.S. Department of Agriculture's Economic Research Service, accessed May 2007, available at www.ers.usda.gov/Data/Food-Consumption.

59 *Older chicken and turkey breeds were spangled, barred, white, black, red, buff, blue, and "Columbian":* Hays, F. & Klein, G., *Poultry Breeding Applied*, Poultry-Dairy Publishing Co., pp. 30–33 (1943).

60 *artificial insemination for chickens had first been developed:* Hays & Klein, pp. 147 & 176.

60 *"the best tasting turkey in America.":* Burros, M., "The Hunt for a Truly Grand Turkey: One that Nature Built," *New York Times*, November 21, 2001.

61 *a typical building for meat chickens produces more than 200 tons of poultry litter:* Website of Kentucky Enrichment, available at: http://www.kentuckyenrichment.com/pages/byproduct/litter_feed_overview.html

Chapter 4: Facing off with Big Ag

64 *they were directly and indirectly contaminating groundwater:* Groundwater is not explicitly covered by the Federal Clean Water Act. However, much groundwater contamination ends up polluting surface waters. Such surface water pollution is arguably covered by the Act.

79 *people living near hog factories suffer higher than normal levels of depression, anxiety, nausea, and sore throats:* Wing, S., & Wolf, S., "Intensive Livestock Operations, Health and Quality of Life Among Eastern North Carolina Residents, University of North Carolina, School of Public Health, *Environmental Health Perspectives*, vol. 108, no. 3 (March 2000).

79 *industrial hog facilities were disproportionately located in communities with high levels of poverty:* Wing, S., et al., "Environmental Injustice in North Carolina's Hog Industry," University of North Carolina, School of Public Health, *Environmental Health Perspectives*, vol. 108, no. 3 (March 2000).

79 *She had written her Ph.D. thesis . . . about human health problems related to hog confinement facilities:* Okun, M., "Human Health Issues Associated with the Hog Industry," University of North Carolina, School of Public Health (January 1999).

Chapter 5: It's a Pig's Life

96 *in the early 1960s, more than 75,000 family hog farmers in North Carolina were raising about a million hogs a year:* USDA's National Agricultural Statistics Service, accessed August 2007.

99 *air laden with ammonia, hydrogen sulfide, dust, viruses, bacteria, and endotoxins:* Endotoxins are toxins found in certain bacteria that are released when the bacteria break down or die. Studies show they are found at high levels in animal confinement operations. For more on air contaminants in swine and poultry confinement buildings, see: Clark, S., et al., "Airborne Bacteria, Endotoxin and Fungi Dust in Poultry and Swine Confinement Buildings," *American Industrial Hygiene Association Journal*, vol. 44, no. 7, pp. 537–41 (1983).

99 *97 percent of hogs in finishing operations are continuously given antibiotics:* "Swine '95: Grower/Finisher, Part II: Reference of US Grower/Finisher Health and Management Practices," a publication of USDA, cited by, Mellon, M., et al., *Hogging It: Estimates of Antimicrobial Abuse in Livestock*, a report of the Union of Concerned Scientists, p. 39, 2001.

99 *"Encourage sufficient exercise...":* Anderson, A.L., *Swine Management*, J.B. Lippincott Company, p. 188 (1950).

99 *"it is essential for the sow...":* Dawson, H.C., *The Hog Book*, Sanders Publishing Company, p. 129 (1913).

108 *pigs are said to communicate with their tails:* Hahn Niman, N., "The Unkindest Cut," *New York Times*, March 7, 2005.

109 *Tail biting is common among pigs in confinement:* "Modern slaughter pig production often takes place in fully slatted pens with high animal densities. Such an environment may predispose to aberrant behavior such as tail biting and elevated aggressive activity." Simonsen, H., "Effect of Early Rearing Environment and Tail Docking on Later Behavior and Production of Fattening Pigs," *Acta Agricultural Scand. Sect. A. Animal Science*, vol. 45, p. 139 (1995).

109 *"tail docking was associated with a three-fold increase...":* Moinard, C. et al., "A Case Control Study of On-farm Risk Factors for Tail Biting in Pigs," *Applied Animal Behavior Science*, vol. 2016 (2003).

109 *"A behaviorally appropriate environment...":* "Humane Husbandry Criteria for Pigs," Animal Welfare Institute, sec. 11(b), available at: www.awionline.org/farm/standards/humane.

109 *docking pigs' tails has been prohibited in the European Union since 1991:* Stevenson, P., "Animal Welfare Issues in the Intensive Farming of Pigs in the European Union," Conference paper, December 16, 2000.

111 *The article described a Washington state cattle slaughterhouse:* Warrick, J., "They Die Piece by Piece," *Washington Post*, April 10, 2001.

113 *painting an unflattering portrait of the facility:* Petersen, M., "Indians Now Disdain a Farm Once Hailed for Giving Tribe Jobs," *New York Times*, November 15, 2003.

120 *Niman was selling ... to fine restaurants in the Bay Area under the name "Niman Ranch,":* Bill Niman was joined in the enterprise for a time by journalist Orville Schell. During those years, the company was known as "Niman-Schell Ranch." In the mid-1990s, Niman bought out Schell's interest in the company and the name became "Niman Ranch." Orville Schell went on to become Dean of the University of California at Berkeley Graduate School of Journalism.

Chapter 6: A Door Closed, a Door Opened

128 *the second Hog Summit ... would take place at the Surf Ballroom:* The Surf Ballroom is known to many as the last place where Buddy Holly, Ritchie

Valens and Big Bopper all played just before dying in a plane crash on February 2, 1959.

Chapter 7: Beef, the Most (Unfairly) Maligned of Meats

136 *Cattle belong to the zoological order Artiodactyla:* Hickman, C., *Integrated Principles of Zoology*, 7th Ed., p. 662, Times Mirror / Mosby College Publishing (1984).

136 *"the most important step ever taken by man in exploitation of the animal world":* Davis, P., & Dent, A., *Animals that Changed the World: The Story of the Domestication of Wild Animals* Crowell-Collier Press, p. 66 (1968).

136 *bovine domestication may have begun as far back as eleven thousand years ago:* The ancient Near East encompasses roughly the same area as what is now called the Middle East. Data regarding cattle domestication from: Beja-Pereria, A., et al., "The Origin of European Cattle: Evidence from Modern and Ancient DNA," *Proceedings of the National Academy of Sciences*, May 11, 2006, available at: http://www.pnas.org/cgi/content/abstract/0509210103v1.

136 *Cattle first arrived in the Americas in 1493:* Snapp, Roscoe P., *Beef Cattle: Their Feeding and Management in the Corn Belt States*, 3d Ed., John Wiley & Sons, Inc., p. 3 (1939).

136 *original seed stock for Latin America's cattle:* Wellman, P., *The Trampling Herd: The Story of the Cattle Range in America*, University of Nebraska Press, pp. 14–17 (1939).

137 *"The long grazing season, mild winters ..."* Snapp, R. P., p. 5, *citing* USDA Yearbook, 1921, p. 232.

137 *"Prior to the Civil War, the Ohio and upper Mississippi Valley ...":* Snapp, R.P., p. 16.

137 *as early as the Middle Ages cattle were fed* exclusively *"hay and corn":* Whitlock, Ralph, *A Short History of Farming in Britain,* John Baker Ltd. (1965).

137 *"Grass was still abundant and relatively cheap ...":* Snapp, R.P., p. 16.

137 *supplemented with a liberal allowance of shock corn:* Shock corn is dried, harvested, and bundled in the field then fed while still whole and on the stalk.

138 *"As these western cattle proved highly satisfactory ...":* Snapp, R.P., p. 16.

138 *27 million cattle dotted the country:* Snapp, R.P., p. 13, *citing* USDA data.

138 *Although they were fed hay and grain at certain life-stages and times:* Note, too, that already at the dawn of the twentieth century, USDA's agricultural research stations were studying feeding grain to cattle for fattening. "Experiments at the Iowa Station were concerned with light, medium and heavy grain feeding in 1904." Riggs, J., "Fifty Years of Progress in Beef Cattle Nutrition," *Journal of Animal Science*, vol. 17, p. 987 (1958).

140 *a 1998 report for the World Bank and United Nations:* Mearns, R., p. 7; See also, Elpel, T., "The American Sahara, the New Desert Beneath Our Feet,"

available at: www.wildflowers-and-weeds.com/sahara.htm, which argues: "Cows are not the problem. The problem is our management practices. . . . Adding to the ironies of desertification is the fact that many environmentally conscientious people adamantly despise cows and are willing to do anything to remove them from public lands—precisely because these smelly beasts trample, manure, destroy everything around them. Environmentalists are even more conservative than ranchers to embrace the new paradigm of range ecology, because to do so implies an endorsement of cattle on public lands. They do not want to hear that cows can be good for the land, and they try to forget the idea as quickly as they hear. In short, the environmental community is unwittingly fueling North America's greatest environmental disaster."

141 *good herd management can prevent the erosion and land degradation:* See: Website of the Natural Resources Conservation Service, Grazing Lands Issue Brief, November 1995, available at: www.nrcs.usda.gov/technical/nri/pubs/ib6text.html, which notes: "Fortunately, grazing lands can be maintained in a healthy state with grazing, and properly managed grazing can enhance ecosystem health."; and Mearns, R., *Livestock and Environment: Potential for Complementarity*, September 1996, available at: www.fao.org/DOCREP/W5256T/w5256t02.htm , which states: "A wealth of evidence exists to support the view that light or moderate grazing by livestock increases rangeland productivity in many grazing systems."

141 *Properly timed grazing triggers beneficial biological processes that improve soil:* Manske, L., "Effects of Grazing Management Treatments on Rangeland Vegetation," North Dakota State University, website of Dickinson Research Extension Center, 2004, available at: www.ag.ndsu.nodak.edu/dickinso/research/2003/range03c.htm.

141 *an estimated 10 million elk:* Website of Rocky Mountain Elk Foundation, www.rmef.org, accessed December 2007.

141 *and 30 to 75 million bison:* Website of Fish and Wildlife Service, www.fws.gov, accessed December 2007.

141 *"The moving multitude . . . darkened the whole plains.":* Journals of Lewis and Clark, according to website of Fish and Wildlife Service, accessed December 2007.

141 *40 million mature dairy and breeding beef cows:* USDA data for 2007, accessed January 2008.

141 *"When grazers are removed . . .":* Curry, J.P., (professor of Environmental Resource Management, University College, Dublin), *Grassland Invertebrates* (1993).

141 *"grazing by large herbivores is fundamental . . .":* Harnett, D., Kansas State University, presented at the 1999 Society for Range Management Meeting, February 1999, website of Ecological Society of America, available at: www.esa.org/science_resources/publications/purePraireLeague.php.

142 *"Grass prevents erosion by binding the soil.":* Rogers, J., & Feiss, P.G., *People and*

the Earth: Basic Issues in the Sustainability of Resources and Environment, Cambridge University Press, p. 63 (1998).

142 *"replacement of grasslands with cropland . . ."*: Rogers & Feiss, p. 63, (1998).

142 *Even wild areas of brush can fail:* Reynolds, H., "Effects of Burning on a Desert Grass-Shrub Range in Southern Arizona," *Ecology,* vol. 37, October 1956, which notes: "Shrub encroachment thus leads to more flash flood runoffs and increased erosion."

142 *Large grazing animals also disperse seeds:* Mearns, R., *Livestock and Environment: Potential for Complementarity,* report to FAO and World Bank, December, 1996. See also, Janzen, D., "Dispersal of Small Seeds by Big Herbivores, Foliage is the Fruit," *The American Naturalist,* March 1984 (abstract). "Many species of herbs (including grasses) have some of their seeds dispersed by the large grazing mammals that consume the seeds along with foliage. This is an interaction that has probably been occurring for many millions of years. It should result in a very effective kind of seed dispersal to sites newly open for colonization in a wide variety of habitat types."

142 *"Animals can be an effective and economical . . ."*: Andrae, J., "Grazing Impacts of Pasture Composition," University of Georgia College of Agriculture and Environmental Sciences," available at: http://pubs.caes.uga.edu/caespubs/pubcd/B1243.htm.

142 *"permanent pasture for grazing livestock can be an ideal choice for minimizing [water] pollution."*: "Sustainable Farming Systems: Demonstrating Environmental and Economic Performance," study of University of Minnesota et al., June 2001. available at: www.misa.umn.edu/vd/SFSreport.pdf

142 *Like forests . . . grasslands act as carbon sinks:* Boody, G., et al., "Multifunctional Agriculture in the United States," *Bioscience,* vol. 55, p. 32 (2005).

142–43 *"[t]here is growing evidence that both cattle ranching and pastoralism . . ."*: *Livestock's Long Shadow: Environmental Issues and Options,* Report to the U.N. Food and Agriculture Organization, p. 254, 2007, available at: http://www.fao.org/docrep/010/a0701e/a0701e00.htm.

143 *"Grazing seven months a year and feeding hay . . ."*: Sutherly, B., "Bob Evans Wary of Production Driven Farms, Sausage Baron Touts Year-Round Grazing," *Dayton Daily News,* December 4, 2002.

144 *Feedlots then keep cattle for 150 to 270 days:* Environmental Protection Agency Effluent Guidelines for Concentrated Animal Feeding operations, p. 4–89; document available at: http://epa.gov/guide/cafo/pdf.

145 *the rising costs of owning or leasing real estate:* In the late 1940s, land prices steadily rose to the point that "ranch farmers commonly had from $150 to $500 invested in land for each head of cattle they carried," which was "at least a 50 percent inflation in the true grazing value of the land," according to John T. Schlebecker, curator of the Smithsonian's Division on Agriculture and Mining, in *Cattle Raising on the Plains, 1900–1961,* University of Nebraska Press, p. 191 (1963). The ever-escalating price of land made it more expensive every year to keep animals on pasture. Land cost remains one of

the primary challenges for American farmers. According to USDA figures, in the past decade, the average value of an acre of farm land has more than doubled, going from $489 per acre to $1,160 per acre as of August 2007.

146 *In the early twentieth century, young cattle were still being grown to full maturity:* Snapp, R.P., p. 231.

146 *by the 1930s . . . many two-year-olds found themselves in feedlots:* Snapp, R.P., p. 231.

147 *After a 1954 study reported inexpensive, dramatic increases in feedlot gains, hormone use spread:* Riggs, J., p. 1000, *citing* Burroughs, et al., *Science*, vol. 120, p. 66 (1954).

147–48 *Already by 1958, an estimated 70 percent of cattle in feedlots were being administered synthetic hormones:* Schlebecker, J., *Cattle Raising on the Plains, 1900– 1961,* University of Nebraska Press, p. 231 (1963).

148 *"No compound had as much impact on the beef cattle industry . . .":* Riggs, J., p. 1000.

148 *Today, hormones for beef animals are even more popular:* Hanrahan, C., "The European Union's Ban on Hormone-Treated Meat," *Congressional Research Service Report for Congress,* December 19, 2000, (accessed January 7, 2008), available at: www.ncseonline.org/nle/crsreports/agriculture/ag-63.cfm.

148 *an EU scientific committee has twice undertaken thorough reviews:* opinion of the Scientific Committee on Veterinary Measures Relating to Public Health on review of previous SCVPH opinions of 30 April 1999 and 3 May 2000 on the potential risks to human health from hormone residues in bovine meat and meat products, (accessed January 7, 2008), available at: www. ec.europa.eu/food/fs/sc/scv/out50_en.pdf.

148 *humans and wildlife are at real risk of exposure to hormones generated by cattle feedlots:* opinion of the Scientific Panel on Contaminants in the Food Chain on a request from the European Commission Related to Hormone Residues in Bovine Meat and Meat Products, adopted on 12 June 2007.

148 *American research linking . . . cattle feedlot runoff to endocrine disruption in fish:* Orlando, E., "Endocrine Disrupting Effect of Cattle Feedlot Effluent on an Aquatic Sentinel Species," March 2004, available at: http:// www.ehponline.org/docs/2003/6591/abstract.html.

148–49 *the U.S. Food and Drug Administration approved the antibiotic aureomycin:* Schlebecker, J., p. 231.

149 *Research documented that it and other antibiotics sped cattle gain:* Schlebecker, J., p. 229.

149 *FDA also approved adding tranquilizers:* Schlebecker, J., p. 231.

149 *"[T]he addition of antibiotics to feed promised to be . . .":* Schlebecker, J., p. 229.

149 *"Scientists continued to try various combinations . . .":* Schlebecker, J., p. 231.

149 *"At the rate of discovery and use in the fifties . . .":* Schlebecker, J., p. 231.

149 *urea . . . caused "varying degrees of kidney congestion" and was fatal.:* Riggs, J., pp. 989–990.

149 *Yet a 1951 surplus of fats:* Riggs, J., p. 986.

150 *experimental stations examined the use of "both edible and inedible animal fats,":* Riggs, J., p. 986–987.

150 *in April 1985, a frightening sickness surfaced in the United Kingdom:* Schwartz, M., *How the Cows Turned Mad*, (translated by Edward Schneider), University of California Press, pp. 109, 144 (2003).

150 *with two million cattle infected in Britain alone:* Arthur, C., "Scientists double estimate of BSE-infected cattle," *The Independent*, October 10, 2002.

150 *link between the cattle disease and the debilitating and fatal human illness, variant Creutzfeldt-Jakob disease:* Centers for Disease Control: http://www.cdc.gov/ncidod/dvrd/vcjd/epidemiology.htm.

151 *cattle on the range studiously avoid eating any kind of manure:* I have often observed on our ranch that cattle—in striking contrast to dogs—avoid any kind of manure, whether from cattle or other animals. See also: Halverson, M., *Farm Animal Health*, a report for the Minnesota Planning Agency Environmental Quality Board, April 23, 2001, p. 110, who notes of cattle: "Herbage contaminated with feces is usually rejected."

151 *Experiments especially flourished in the 1960s and 1970s:* Such experiments with manure feeding are summarized in a United Nations Food and Agriculture Report, available at: http://www.smallstock.info/reference/FAO/004/X6518E/X6518E03.htm.

151 *"It can be concluded . . . that manure derived from cattle . . .":* From FAO report, Sec. 2.5 Feeding Cattle Waste.

151 *"These are important benefits that inorganic fertilizer does not offer.":* Cogger, C., "Manure on Your Farm: Asset or Liability," USDA, available at: http://www.lpes.org/SmallFarms/3_Manure.pdf.

152 *"Younger cattle generally have more health problems . . ."* Website of Beef Cooperative Research Centre of Australia, available at: http://www.beef.crc.org.au/nutrition/level_1/l4_1.htm.

152 *Mature cattle have complex, four-chambered digestive tracts:* Virginia Tech University article, found at: http://www.ext.vt.edu/pubs/beef/400-010/400-010.html.

152 *digestive tracts of calves are different:* Delton, C, "Calf Rearing–Digestion in the Calf," at http://www.lifestyleblock.co.nz/articles/cattle/11_calf_digestion.htm.

152 *By about three months of age, the calf's digestive tract begins:* Moran, J., *Calf Rearing: A Practical Guide*, CSIRO Publishing (2002).

152 *Acidosis happens frequently when:* Owens, J., "Acidosis in Cattle: A review," *Journal of Animal Science*, vol. 76, pp. 275–286, January 1998.

152 *When cattle can't rid themselves of the gas:* For discussion of bloat, see: http://www.cattletoday.com/archive/2003/March/CT260.shtml.

152 *as a result of an acidosis episode:* "Animal Health in Beef Cattle Feedlots," Queensland Department of Primary Industries and Fisheries (accessed January 2008), available at: http://www2.dpi.qld.gov.au/health/3548.html.

153 *feedlots add a category of antimicrobials called "ionophores":* Shaver, R., "Feeding to Minimize Acidosis and Laminitis in Dairy Cattle," University of Wisconsin extension website, (accessed January, 2008) available at: http://www.extension.org/pages/Feeding_to_Minimize_Acidosis_and_Laminitis_in_Dairy_Cattle.

153 *"Feeding higher amounts of dietary roughage . . .":* Owens, J., January 1998.

153 *PEM, too, is associated with acidosis:* "Animal Health in Beef Cattle Feedlots," Queensland Department of Primary Industries and Fisheries (accessed January 2008), available at: http://www2.dpi.qld.gov.au/health/3548.html.

153 *there's the problem of coccidiosis:* Kirkpatrick, J., DVM, "Coccidiosis in Cattle," Oklahoma Cooperative Extension Service website, pub. No. F-9129, available at: http://osuextra.okstate.edu/pdfs/F-9129web.pdf.

153 *"[T]he younger the animal, the more susceptible . . .":* "Coccidiosis in Cattle," Website of Cornell University Co-operative Extension, (accessed January 2008), available at: http://www.nwnyteam.org/AgFocus2007/Nov/Coccidiosis1.htm.

153 *Coccidiosis "occurs commonly in overcrowded conditions . . .":* Kirkpatrick, at: http://osuextra.okstate.edu/pdfs/F-9129web.pdf.

154 *The guide recommends mixing drugs:* (Emphasis my own). From online Merck Veterinary manual, available at: http://www.merckvetmanual.com/mvm/index.jsp?cfile=htm/bc/21202.htm.

154 *The most recent official census of agriculture showed:* EPA CAFO Effluent Guidelines, p. 4–92, available at: http://epa.gov/guide/cafo/pdf/DDChapters1-4.pdf.

154 *"The eye of the master . . .":* Henry, W., & Morrison, F., *Feeds and Feeding: A handbook for the student and stockman* (1927).

154 *Researchers . . . randomly sampled 200 packages of ground meat:* White, D., et al, "The Isolation of Antibiotic-Resistant Salmonella from Retail Ground Meats," *The New England Journal of Medicine*, vol. 345, no. 16 (October 2001).

155 *about 21,000 pounds per feedlot steer per year:* Note that this assumes an average live weight of 1,000 pounds. EPA Risk Assessment, 2004, p. 15.

155 *83 percent dispose of their waste through land application:* EPA CAFO Effluent Guidelines, p. 4–100, available at: http://epa.gov/guide/cafo/pdf/DDChapters1-4.pdf.

155 *"Traditional means of using manure are not adequate . . .":* U.S. EPA Risk Assessment for Concentrated Animal Feeding Operations, May 2004, available at: http://www.epa.gov/nrmrl/pubs/600r04042/600r04042.pdf.

155 *majority of large feedlots—92 percent—have no land:* EPA CAFO Effluent Guidelines, Table 4-82.

155 *Feedlots with fewer than a thousand animals, however, fare much better:* EPA CAFO Effluent Guidelines, Table 4-82.

155 *University of Nebraska expects to find that hormones used on feedlot cattle:* Information on this on-going study is available at: http://cfpub.epa.gov/ncer_

abstracts/index.cfm/fuseaction/display.abstractDetail/abstract/8426/
report/0.

156 *"Although the hormone content of waste has not been systemically studied . . .":* EPA
Risk Assessment, 2004, p. 42, available at: http://www.epa.gov/nrmrl/
pubs/600r04042/600r04042.pdf.

156 *"This waste may come in contact with humans . . .":* EPA Risk Assessment, 2004,
p. 36.

156 *feedlots also cause serious odor and air pollution:* EPA CAFO Effluent Guidelines, p. 13-6; and Jacobson, L., et al., "Air Emissions from Animal
Production Buildings," 2003, Available at: http://www.isah-soc.org/
documents/mainspeakers/18%20JacobsonUSA.doc.

156 *"There is evidence that fly populations . . .":* "Management of Nuisance Fly
Populations on Cattle Feedlots," website of Queensland Department of
Primary Industries and Fisheries, (accessed January 2008), available at:
http://www2.dpi.qld.gov.au/environment/8099.html.

158 *freeze branding, a technique that uses extreme cold:* Carpenter, J., "Identification Methods for Beef Cattle," pub. No. 7015, *Co-operative Extension Service
Cattlemen's Library.*

159 *Grazing cattle on these tainted fields:* "Sublethal doses of nitrate may induce
abortion because nitrate readily crosses the placenta and causes fetal
methemoglobinemia and death." Knight, et al, Plants Causing Sudden
Death, August 20, 2002, p. 5. Available at: www.ivis.org/special–books/
knight/chapt1/ivis.pdf.

159 *"Toxic hay is killing cattle across the state . . .":* Gregory, A., "Whole Herds
Dying of Toxic Hay; Officials Urge Farmers to Test for Nitrate Levels,"
The Charlotte Observer, February 21, 2003.

159 *Forages with nitrate between 1 and 1.5 percent:* Poore, M., et al., "Nitrate Management in Beef Cattle Production Systems," West Virginia University
Extension Service, p. 5, August 1999.

159 *found many cases of forages with high nitrate concentrations in hays:* "Excessive
N-fertilization rates may cause high levels of nitrate in harvested forage
which can be harmful to ruminant livestock. Forages with nitrate levels
below 0.4% NO3-N are considered safe for ruminant consumption but
many instances of nitrate concentration greater than 0.4% were reported
in 1997, primarily in hays harvested from swine production facilities."
Ranells, N., et al., "Minimizing Impacts of Swine Lagoon Effluent on
Quality of Receiving Waters Through Efficient Crop Utilization of Nutrients," accessed May 2004, from: www.cals.ncsu.edu.

160 *"[I]n recent years, nitrate problems are becoming common . . .":* Poore, p. 1.

160 *Columbia University has estimated:* www.medicalecology.org/food/livestock/f_
pork.html.

160 *"can cause disease in livestock, including Coccidiosis . . .":* EPA's Producers' Compliance Guide for CAFOs, November 2003, p. 8.

160 *bacteria in the* Brucella *family:* www.medicalecology.org.

Chapter 8: The Un-sacred (Milk) Cow

175 *Bovines in the wild spend most of their waking hours:* Halverson, M., *Farm Animal Health and Well-Being*, a report prepared for the Minnesota Planning Agency, Environmental Quality Board, p. 110, April 23, 2001.

176 *Laying hens ... are simply thrown, still alive, into rendering processors:* Duncan, I., "Animal Welfare Issues in the Poultry Industry," *Journal of Applied Animal Welfare Science*, vol. 4, no. 3 (2001), abstract available at: http://www.lea-online.com/doi/abs/10.1207/S15327604JAWS0403_04.

176 *dairy cattle ... once made up ... half of U.S. beef:* Eckles, C., & Anthony, E., *Dairy Cattle and Milk Production*, 1st Ed., Macmillan Co., p. 15 (1956).

176 *"veal calves are merely a by-product ...":* Eckles & Anthony, p. 257.

177 *a modern form of grazing (often called "rotational grazing"):* The term "rotational grazing" was coined by Andre Voisin in his 1959 work, *Grass Productivity*. The practice is also commonly referred to as "intensive grazing," and "management intensive grazing."

178 *about 4.5 million male calves were born on U.S. dairies:* Data from 1999. Millman, S., "Welfare of surplus calves in the dairy industry," abstract, *Journal of Animal Science*, vol. 78, Suppl. 1, 2000, available at: http://www.asas.org/jas/00meet/part2.pdf.

178–79 *For the first half of the twentieth century, Wisconsin's dairy farming:* Specifically, Wisconsin had 178,000 farms in 1910; 200,000 in 1935; 174,000 in 1950. USDA data, accessed January 2008, available at: http://www.nass.usda.gov/QuickStats/PullData_US.jsp.

179 *the state lost more than half of all farms:* Specifically, in 1959 Wisconsin had 131,215 farms (of which 103,143 were dairy farms). By 1997, there were 65,602 farms (of which 22,576 were dairies). "Numbers of Farms and Dairy Farms in Wisconsin, 1959–1997," Program on Agricultural Technology Studies, accessed January 2008, available at: http://www.pats.wisc.edu/daigra20.html.

179 *In 1945, the average Wisconsin herd had just 15 cows:* Palmer, R., "The Cumulative Impact of Dairy Industry Restructuring," University of Wisconsin, available at: http://www.wisc.edu/dysci/uwex/mgmt/pubs/Financial-Analysis.html.

179 *By 1975, the average had risen to 34:* Barham, B., et al., "Expansion, Modernization, and Specialization in the Wisconsin Dairy Industry," University of Wisconsin, March 2005, available at: http://www.pats.wisc.edu/pdf%20documents/Modernization,_Expeansion_WI_Dairy.pdf.

179 *"has several herds of over 1,000 cows and is building ...":* Palmer, R., "The Cumulative Impact of Dairy Industry Restructuring."

179 *Figures from 1982 show:* Stanton, B., "Regional Conflicts in Dairy Policy," Table 2, Cornell University, available at: http://ageconsearch.umn.edu/bitstream/123456789/18263/1/ar830129.pdf.

179 *Even in 1994, not a single dairy in the state:* USDA data, accessed March 2008.

179 *By 1994, California had 1,800 dairies:* USDA data, accessed March 2008; Stanton, B., "Regional Conflicts in Dairy Policy," Table 2, Cornell University, available at: http://ageconsearch.umn.edu/bitstream/123456789/18263/1/ar830129.pdf.

180 *"In the colonial days in America the cow was seldom . . .":* Eckles & Anthony, p. 3.

180 *"Dairy cattle are more widely distributed on farms . . .":* Petersen, W.E., *Dairy Science*, 1st Ed., J.P. Lippincott Co., p. 22 (1939).

180 *"in the decades following WWII farmers found . . .":* Winsten, J.R., et al., "Differentiated Dairy Grazing Intensity in the Northeast," Pennsylvania State University, *Journal of Dairy Science*, vol. 83, p. 836 (2000).

181 *"The maximum production reached at this season . . .":* Eckles & Anthony, p. 534.

181 *"cows should be turned outside . . .":* Eckles & Anthony, p. 333.

181 *"Dairy cows when kept confined in the barn . . .":* Eckles & Anthony, p. 334.

181 *"No satisfactory plan has been devised . . .":* Eckles & Anthony, p. 562.

182 *In California's San Joaquin Valley, a region that contains 2.5 million dairy cows:* "California air pollution control district updates dairy volatile organic compound emissions estimate," Fresno, CA, August 1, 2005, available at: http://www.caprep.com/0805001.htm.

183 *"The dairy calf is almost always reared by hand . . .":* Eckles & Anthony, p. 230.

184 *It had become normal procedure to separate calves from their mothers:* Eckles & Anthony, pp. 233–240.

184 *These days, female calves are generally fed:* Halverson, M., April 23, 2001, p. 113.

184 *"Approximately 10 pounds of milk are required . . .":* Eckles & Anthony, p. 258.

185 *the Parises were indeed running the farm with a lot less machinery:* This is a typical experience. A Michigan State study found: "Per cow investment for facilities, machinery and equipment . . . is $671 per cow for grazing and $1,895 per cow for confinement." Conner, D., et al., "Opportunities in Grazing Dairy Farms: Assessing Future Options," C. S. Mott Group for Sustainable Agriculture at Michigan State University, July 2, 2007, available at: http://www.mottgroup.msu.edu/Portals/mottgroup/Opportunities%20in%20Grazing%20Dairy%20Farms.pdf.

185 *they take care of most of their own manure:* A University of Missouri study noted: "Research . . . demonstrated dairy animals in a [grazing] system deposited over 80 percent of their feces and urine on pasture." Hamilton, S., et al., "Economics of a Missouri Pasture-based Dairy, Can a Small Farm Survive?" University of Missouri Extension, (417) 637-2112, available at: http://agebb.missouri.edu/modbu/archives/v2n2/survive.pdf.

186 *They are by far the most common dairy breed in the country:* Available at: http://aged.calpoly.edu/AgEd410/Presentations/DairyBreeds.ppt.

186 *"Pasture-based livestock farms can perform significantly better . . .":* Summarized in "Farm Flashes," July-August 2007, p. 21, a publication of Cornell University, available at: http://counties.cce.cornell.edu/oneida/Publica-

tions/Farm%20Flash%20JulyAugust%2007.pdf . Full study at: Conner, D., et al., "Opportunities in Grazing Dairy Farms: Assessing Future Options," C. S. Mott Group for Sustainable Agriculture at Michigan State University, July 2, 2007, available at: http://www.mottgroup.msu.edu/Portals/mottgroup/Opportunities%20in%20Grazing%20Dairy%20Farms.pdf.

187 *"average or better management . . .":* Kriegl, T., "Wisconsin Grazing Dairy Profitability Analysis," University of Wisconsin Center For Dairy Profitability, Food Animal Production systems: Issues and Challenges Conference, Lansing, Michigan, (May 2000) available at: http://cdp.wisc.edu/pdf/gzprofitncr4yr.pdf.

187 *University of Missouri analysis:* Hamilton, S., et al.

188 *" . . . grazing dairies may present more accessible start-up opportunities . . .":* Conner, D., et al., "Opportunities in Grazing Dairy Farms: Assessing Future Options," 2007.

189 *grazing dairy cows produce higher levels of the beneficial fatty acids:* Clancy, K., Union of Concerned Scientists Report, *Greener Pastures*, pp. 2, 42-3, March 2006, available at: http://www.ucsusa.org/food_and_environment/sustainable_food/greener-pastures.html.

190 *Minnesota has a strong tradition of family dairy farms:* Stanton, B., "Regional Conflicts in Dairy Policy," Table 2, Cornell University, available at: http://ageconsearch.umn.edu/bitstream/123456789/18263/1/ar830129.pdf.

192 *35 to 56 cases of lameness anually.* Halverson, M., p. 120.

192 *"The main walking surfaces for cows play a major role . . .":* Halverson, M. p. 120, citing Rushen, J., "The Welfare of the High Producing Animal," Presentation to the Sustainable Animal Agriculture Series, September 2000.

192 *Over time, this strained position is crippling:* Halverson, M., p. 118, citing, Webster 1995.

192 *"Heavy body weight increases the likelihood . . .":* Halverson, M., p. 114.

192 *the cows' tails had all been cut off to stubs:* Hahn Niman, N., "The Unkindest Cut," *New York Times*, March 7, 2005.

193 *researchers noted that when it comes to tail docking:* The study also concluded that there were no benefits to human health: "This study did not identify any differences in udder or leg hygiene or milk quality that could be attributed to tail docking." Schreiner, D.A. & Ruegg, P.L., "Effects of Tail Docking on Milk Quality and Cow Cleanliness," *Journal of Dairy Science*, vol. 85, pp. 2503–2511 (2002). See also: "On the basis of available peer reviewed studies and governmental sponsored research, we conclude that there is ample evidence . . . that there is no benefit to tail docking in dairy cattle. Presently, there are no apparent animal health, welfare, or human health justifications to support this practice. Until evidence emerges that tail docking has benefits to animal well-being, health, or public health, the routine practice of tail docking should be discouraged." *Journal of the*

American Veterinary Medical Association, vol. 220, no. 9, pp. 1298–1303 (May 1, 2002).

196 *As early as the mid-1950s, calves were being fed blood meal:* Eckles & Anthony, p. 241.

197 *jumped to 22 percent by 1999 and continue to spread:* Recent data show that managed grazing is practiced on about 23 percent of Wisconsin dairy farms. Another 21 percent use mixed feed (partial but not total grazing), while 56 percent are stored feed [total confinement] farms. "Grazing in the Dairy State," Report of University of Wisconsin Center for Integrated Agricultural Systems, January 2006, available at: http://www.cias.wisc.edu/pdf/statussm.pdf.

197 *well-characterized breeds of domestic cattle by 2500 B.C.:* Davis & Dent, p. 61.

197 *distinct types among cattle were mostly formed by natural conditions:* Eckles & Anthony, p. 22.

197 *"a great interest was aroused in England in the improvement . . .":* Eckles & Anthony, p. 22.

198 *With the advent of artificial insemination:* By 1954, over a million dairy cows in more than 600,000 herds in the U.S. were artificially inseminated. Eckles & Anthony, p. 303. (I have not yet met a single dairy farmer who uses a live bull to breed his herd.)

198 *Devon, which was developed in Devonshire, England:* Eckles & Anthony, p. 96.

198 *pounds of milk:* The dairy industry speaks of milk production in pounds, not gallons. One gallon of milk is 8.6 lbs.

198 *produced as much as 5,000 pounds of milk annually:* Eckles & Anthony, p. 96.

198 *"in 1906 the Brown Swiss breeders . . .":* Eckles & Anthony, p. 89.

198 *American versions had been bread to a more "extreme dairy" body:* Eckles & Anthony, pp. 44–84.

199 *"Various classifications of cattle have been made . . .":* Eckles & Anthony, p. 24.

199 *frequent problems for both mother cows and newborns:* Several farmers have told me that Holsteins caused them more problems in calving than crossbred cows or other breeds. A 2002 survey by University of Wisconsin found that farmers gave Holsteins the lowest score for "calving ease" compared to various crossbreeds. Weigel, K. & Barlass, K., "Producers Who've Tried Crossbreeding Share Their Experiences" University of Wisconsin, October 22, 2002, available at: http://www.wisc.edu/dysci/uwex/genetics/pubs/Crossbreeding_Survey_Oct02.pdf.

199 *abuses seem to happen at plants that specialize in the slaughter of old dairy cows:* For example, on January 30, 2008, the Humane Society of the United States unveiled an undercover investigation of a dairy cow slaughter plant in Chino, California. Temple Grandin, a renowned animal agriculture expert, called images captured in the investigation "one of the worst animal abuse videos I have ever viewed." The result was a recall of 143 million pounds of ground beef, the largest in U.S. history.

200 *their producers marketed them with a vengeance:* Petersen, pp. 40–1.

200 *"The keeping of dairy animals was the greatest factor . . .":* Eckles & Anthony, citing Dr. Elmer V. McCollum, pp. 1, 7.

200 *Breeding and feeding were the two primary tools in pushing up her production:* Eckles & Anthony, p. 29.

200 *In 1850 each milked cow had fed just over three persons:* Eckles & Anthony, p. 5.

200 *Americans were consuming an average of three pounds of cheese:* USDA data (accessed February 2008).

200 *a cow produced an average of 2,902 pounds of milk per year:* Petersen, W.E., *Dairy Science*, p. 10 (1939).

200 *By 1950 each dairy cow was being asked to feed six:* Eckles & Anthony, p. 5.

201 *average had risen to 7,300 pounds:* Eckles & Anthony, p. 14.

201 *a billion pounds of cheese:* USDA data, (accessed February 2008).

201 *up from 1,250 pounds earlier:* Petersen, W.E., p. 106.

201 *19,951 pounds of milk a year:* USDA data (accessed February 2008).

201 *annual cheese production exceeds 9 billion pounds and 'Americans are eating . . . 31 pounds:* USDA data (accessed February 2008).

201 *"Improved genetics accounts for about half . . .":* College Of Agricultural And Life Sciences, University Of Wisconsin-Madison, Cooperative Extension, University Of Wisconsin-Extension, (July 2004), available at: http://aae.wisc.edu/pubs/dairyland/pdf/rd7.pdf.

201 *before even reaching their peak milking age:* Research shows cows reach their peak milking years between six and eight years of age. Halverson, M., p. 116, citing Blakely and Bade 1979.

201 *"make animal-marine protein by-products particularly attractive . . .":* Shaver, R., "By-Product Feedstuffs in Dairy Cattle Diets in the Upper Midwest," University of Wisconsin, available at: http://www.wisc.edu/dysci/uwex/nutritn/pubs/ByProducts/ByproductFeedstuffs.html.

202 *"Dairy cattle are also more likely to receive protein supplements . . .":* July 25, 2005 letter from Consumers Union to Agriculture Secretary Johanns, available at: http://www.consumersunion.org/pub/2005/07/002530print.html .

202 *Typical feeding rates:* Shaver, R., "By-Product Feedstuffs in Dairy Cattle Diets in the Upper Midwest," University of Wisconsin, available at: http://www.wisc.edu/dysci/uwex/nutritn/pubs/ByProducts/ByproductFeedstuffs.html.

202 *"coagulated packing house blood which has been dried and ground . . .":* Cullison, A. & Lowrey, R., *Feeds and Feeding*, p. 184 (4th ed. 1987).

202 *Recombinant bovine somatotropin:* One also commonly sees the terms rBGH or BGH which mean "recombinant growth hormone" and "bovine growth hormone." All three terms have essentially the same meaning.

203 *"[Their] use results in high milk yields . . .":* Halverson, M., p. 126, citing (Broom 1993; Kronfeld, et al. 1997; Willeberg 1993, 1997).

203 *"The production of a liberal amount of milk . . .":* Eckles & Anthony, p. 351.

203 *"Defects are common in all breeds.":* Eckles & Anthony, p. 38.

203 *Mastitis, an inflammation of the udder:* Halverson, M., p. 125, (citing Oliver, et al., 2000); and Halverson, M., p. 126, (citing, Faye, et al. 1997; Emanuelson & Oltenacu 1998; Oltenacu & Ekesbo 1994; and Waage, et al. 1998).

203 *milk yields and increased incidences of other maladies:* Halverson, M., p. 114–5.

203 *"They're more sure-footed.":* Sutherly, B., "Bob Evans Wary of Production Driven Farms, Sausage Baron Touts Year-Round Grazing," *Dayton Daily News,* December 4, 2002.

204 *"The capacity and willingness of the U.S. dairy industry to produce milk...":* Coppock, C, (former professor of Dairy Science, Texas A&M), "Selected Features of the U.S. Dairy Industry from 1900 to 2000," February 2000 (accessed January 2008), available at: http://www.coppock.com/carl/writings/History_of_Dairy_Production_From_1900_to_2000.htm.

204 *"The [surplus] problem is not temporary..."* Stanton, B., "Regional Conflicts in Dairy Policy," Cornell University (1983), available at: http://ageconsearch.umn.edu/bitstream/123456789/18263/1/ar830129.pdf.

204 *government even paid dairy farmers to slaughter almost 10 percent:* The slaughter of 10 percent of U.S. dairy cows was called The Dairy Termination Program (or the Dairy Buyout). Coppock, C., p. 9.

204 *the federal government has long purchased cheese, butter, and nonfat dry milk:* Keniston, M., "What is a Dairy Surplus," Dept. of Ag. Economics, Cornell Univ., (January 1990), available at: http://hortmgt.aem.cornell.edu/pdf/smart_marketing/keniston1-90.PDF.

204 *"In 2002, storage costs alone for the powder...":* Highplains/Midwest Ag Journal, May 31, 2005, available at: http://www.hpj.com/archives/2005/jun05/jun6/Governmentrunningoutofsurpl.CFM.

204 *"the dairy cow must be worked to the limits..."* Halverson, M., p. 116.

205 *Zartman is a strong proponent of grass-based farming:* Sutherly, B., "Bob Evans Wary of Production Driven Farms, Sausage Baron Touts Year-Round Grazing," *Dayton Daily News,* December 4, 2002.

Chapter 9: A Bit About Fish

208 *In 1970, fish farming accounted for just 4 percent:* The State of World Fisheries and Aquaculture, a report of the United Nations Food and Agriculture Organization, part 1, 2006, available at: http://www.fao.org/docrep/009/a0699e/A0699E04.htm#4.1.1.

208 *By 2000 it produced 27 percent:* The State of World Fisheries and Aquaculture, 2006.

208 *by 2007, nearly one-half of the seafood consumed in the world:* Speech of U.S. Secretary of Commerce, U.S. Department of Commerce website, March 12, 2007, available at: http://www.commerce.gov/NewsRoom/SecretarySpeeches/PROD01_002825.

208 *"Worldwide, the sector has grown at an average rate . . ."* The State of World Fisheries and Aquaculture, 2006.

209 *A large salmon farm can pour as much untreated liquid waste:* McCarthy, T., "Is Fish Farming Safe?" *Time*, November 17, 2002.

209 *"They're like floating pig farms."* Los Angeles Times, December 9, 2002.

209 *"[T]echnology is now available for farming an increasing range . . ."* Shepard, *Intensive Fish Farming*, p. 15.

209 *U.S. Department of Agriculture aquaculture report from 2004:* Harvey, D., "U.S. Seafood Market Shifts to Aquaculture," April 2004, available at: http://www.ers.usda.gov/AmberWaves/april04/Findings/USSeafood.htm.

210 *"a combination of escalating fuel costs and declining wild fish stocks . . ."* Shepard, pp. 2-3.

210 *fisheries of all major fish species will collapse by the year 2048:* Worm, B., et al., "Impacts of Biodiversity Loss on Ocean Ecosystem Services," *Science*, vol. 314, no. 5800, pp. 787–790, November 3, 2006.

210 *fish meal . . . consists of fish or fish by-products:* Cullison, A. & Lowrey, R., *Feeds and Feeding*, 4th ed., Prentice-Hall, Inc., p. 183, (1987).

211 *It was being tried for the first time in the United States:* Henry, W. & Morrison, F., *Feeds and Feeding*, 19th ed., Henry-Morrison, p. 186 (1927).

211 *8 percent of fish went into animal feeds in 1948:* Stier, K., "Fish Farming's Growing Dangers," *Time*, September 19, 2007.

211 *fish meal and fish oil became major components of farmed animal feeds:* Hicks, B., "Feeding Fish, Fish Meal And Fish Oil Fulfill Organic Tenets," Organic Seafood Association, Langley, BC, Canada, November 2007, available at: http://www.ams.usda.gov/nosb/MeetingAgendas/Nov2007/OralPresentations/HicksBPanelistPaperFeed11-07.pdf .

211 *By the late 1990s, one-third of the global (wild) fish catch:* Shepard, p. 3.

211 *by 2007, 37 percent of global seafood (wild and farmed) was ground up:* Stier, K., "Fish Farming's Growing Dangers," *Time*, September 19, 2007.

212 *"With menhaden in decline, the recovered population of striped bass . . .":* Russell, D., "Managing for the Ecosystem," September 27, 2005, available at: http://www.dickrussell.org/archives.htm.

212 *by turning poultry waste into feed for livestock and farmed fish:* Kentucky Enrichment website, available at: http://www.kentuckyenrichment.com/pages/byproduct/litter_feed_bizplan.html.

212 *In Asia, this is often done with little or no waste treatment:* "U.S. Proposal to Allow Chicken Imports Raises Health Concerns," *Boston Globe*, May 9, 2007.

212 *migratory birds falling ill near several large carp farms:* Ellis, L. & Turner, J., "A China Environmental Health Project Research Brief, Aquaculture and Environmental Health in China," a report of the Woodrow Wilson International Center for Scholars, May 7, 2007, available at: http://www.wilsoncenter.org/topics/docs/fish_may07.pdf.

213 *standards for so-called "organically farmed" fish:* "Organically farmed" fish has been somewhat controversial. As one article put it, "it doesn't deliver the radical difference in production methods consumers have come to expect from other categories of organic food." The article also quoted a key figure in the modern British organic movement as saying: "So-called 'organic salmon' is making a mockery of organic standards." Blythman, J., "Why Organic Salmon is Causing a Nasty Smell," *Observer Food Monthly*, October 22, 2006.

213 *"Slaughter by-products provide essential amino acids . . ."* USDA's National Organic Standards Board (NOSB), Livestock Committee, Invitation for Public Comment on Aquaculture Standards, September 8, 2006, available at: http://www.ams.usda.gov/nosb/lscommRMR/recommendations/Oct06MeetingRecs/LC-FNLInvitationAWG090806.pdf.

213 *"It is the fish farmer's nightmare . . ."* Clover, p. 264.

213 *Bangladesh imports meat and bone meal. The Daily Star*, September 8, 2007, available at: http://www.thedailystar.net/story.php?nid=3040.

213 *Fish feed formulation in China lists the following animal ingredients:* Tacon, A., "Fish Feed Formulation and Production," report of the Food and Agriculture Organization of the United Nations (November 1990).

213 *92 percent of the world's aquaculture: The State of World Fisheries and Aquaculture*, FAO Report, 2006.

213 *China as the globe's leading producer and exporter of seafood:* Barboza, D., "China Says its Seafood is Now Safer and Better," *New York Times*, January 18, 2008.

213 *70 percent of world's aquaculture production: The State of World Fisheries and Aquaculture*, FAO Report, 2006.

213 *"Fish farming was probably first practiced . . .":* Shepard, p. 2.

213 *Traditional Chinese fish farming was based on natural cycles: Time*, November 17, 2002.

214 *China's pollution problem:* Kahn, J., "As China Roars, Pollution Reaches Deadly Extremes," *New York Times*, p. A1, Sunday August 26, 2007.

214 *many Chinese fish farmers douse their fish with drugs and chemicals:* Barboza, D., "China Says its Seafood is Now Safer and Better," *New York Times*, January 18, 2008.

214 *United States, ranking tenth in the world in 2004: The State of World Fisheries and Aquaculture*, FAO Report, 2006.

214 *Almost three-quarters of U.S. aquaculture is in the South:* Data from 1998 Census on Aquaculture, USDA website, available at: http://www.nass.usda.gov/census/census97/aquaculture/aquaculture.htm.

214 *Meat and dairy (except lamb) are mostly of domestic origin:* Some 90 percent of red meat and 97 percent of the dairy products Americans consume are of domestic origin. Data from *Amber Waves of Grain*, report of the Economic Research Service / USDA, (February 2008), available at: http://www.ers.usda.gov/AmberWaves/February08/DataFeature/.

214 *including almost 90 percent of shrimp:* "Prepare for Jumbo Shrimp Prices," *San Francisco Chronicle*, July 7, 2007.

214 *"[H]igh density fish farms are natural breeding grounds for pathogens . . .":* Dean, C., "Lice in Fish Farms Endanger Wild Salmon, Study Says," *New York Times*, December 14, 2007.

214 *"Microbial infection and multiplication are more likely . . .":* Shepard, p. 19.

214 *"The more aquaculture there is, the more disease . . .":* McCarthy, T., "Is Fish Farming Safe?" *Time*, November 17, 2002.

215 *antibiotics are commonly found in formulated fish feed:* "Occurrence of Antibiotics in Water from Fish Hatcheries," a USGS publication, available at: http://ks.water.usgs.gov/Kansas/pubs/fact-sheets/fs.120-02.pdf.

215 *"with lax standards, raised in ponds where many chemicals are used to keep the shrimp alive.":* Burros, M., "Wegmans Sets Standards for Shrimp," *New York Times*, October 31, 2007.

215 *Drugs given shrimp and fish "pass easily into the surrounding environment . . .":* McCarthy, T., 2002.

215 *In 2007 the U.S. Food and Drug Administration announced:* "US Restrictions on Aquaculture Products from China," Statement by the US Food and Drug Administration, February 2007.

215 *"One of the most damaging organisms . . .":* McCarthy, T., 2002.

215 *"Parasites that breed in fish farms kill so many . . .":* Dean, C., "Lice in Fish Farms Endanger Wild Salmon, Study Says," *New York Times*, December 14, 2007.

215 *the aquaculture industry has caused widespread environmental problems:* "The State of World Fisheries and Aquaculture," United Nations Food and Agriculture Organization Report, part 1, 2006, available at: http://www.fao.org/docrep/009/a0699e/A0699E04.htm#4.1.1.

216 *"long strips of coastline south of Bangkok . . .":* McCarthy, T., 2002.

216 *"Nets are ripped open . . .":* McCarthy, T., 2002.

217 *" 'The Trojan gene effect:' "* "Purdue Scientists: Genetically Modified Fish Could Damage Ecology," *ScienceDaily*, February 24, 2004, available at: http://www.sciencedaily.com/releases/2004/02/040224084432.htm.

217 *the third-most consumed seafood in the United States:* Clover, C., *The End of the Line, How Overfishing is Changing the World and What We Eat*, Ebury Press, p. 255 (2005).

217 *increasing forty-fold over the last two decades:* "Farmed salmon more toxic than wild salmon, study finds," University of Indiana News Room, January 8, 2004, available at: http://newsinfo.iu.edu/news/page/normal/1225.html.

217 *farms actually consume more pounds of fish than they produce:* "Scientists Say Fish Farming Lessens Fish Supply," website of Environmental Defense, Posted: June 28, 2000; Updated: February 5, 2002, available at: http://www.environmentaldefense.org/content.cfm?contentID=1219.

217 *about twenty pounds of feed for every pound of tuna meat:* Stier, K., "Fish Farming's Growing Dangers," *Time*, September 19, 2007.

217 *higher toxin levels occur in farmed salmon:* Clover, p. 250.

217 *farm-raised Atlantic salmon from both Europe and North America:* "Farmed salmon more toxic than wild salmon, study finds," University of Indiana News Room, January 8, 2004, available at: http://newsinfo.iu.edu/news/page/normal/1225.html.

218 *cantaxanthin is added to the feed of salmon:* Mallet, G., *Last Chance to Eat: The Fate of Taste in a Fast Food World*, W.W. Norton & Co., p. 285 (2004); McCarthy, T., "Is Fish Farming Safe?" *Time*, November 17, 2002.

218 *"[I]t is the responsibility of those of us who are fortunate...":* Clover, C., *The End of the Line, How Overfishing is Changing the World and What We Eat*, Ebury Press, p. 251 (2005).

218 *Restaurants, where 70 percent of U.S. seafood consumption takes place:* Stier, K., "Fish Farming's Growing Dangers," *Time*, September 19, 2007.

219 *showed fish displaying signs of fear, suspicion, and doubt:* "Fish Possess Emotions," *New York Times*, November 14, 1915.

219 *"[T]he more time bold and shy fish spent watching...":* "Social Relationships Affect Personality of Fish, Say Experts," University of Liverpool, November 22, 2006, available at: http://www.liv.ac.uk/newsroom/press_releases/2006/11/personality_fish.htm.

219 *fish have distinct, discernable personalities:* "Nothing Fishy about Personality Traits in Animals, Study Finds," *Vancouver Sun*, November 25, 2007.

219 *"If a particular species of fish becomes stressed...":* Shepard, p. 9.

220 *"Just as pain helps an animal that can detect and escape...":* Balcombe, J., *Pleasurable Kingdom: Animals and the Nature of Feeling Good*, Macmillan, p. 189. (2006).

Chapter 10: Finding the Right Foods

224 *The organic standards also provide some assurance about how the animals are housed and handled:* For full text of this regulation, see National Organic Program, § 205.239 Livestock living conditions, available at Code of Federal Regulations, http://ecfr.gpoaccess.gov.

226 *the Certified Humane program has a built-in incentive:* I have several times expressed my concerns about the failings of the Certified Humane program directly to its backers. They indicated they did not see a need to alter the program's standards or its financing.

228 *The color's intensity comes from the xanthophylls:* McGee, H., *On Food and Cooking: The Science and Lore of the Kitchen*, Scribner, p. 75 (2004).

233 *A word here about meat labeled "natural.":* "What is 'Natural' Meat?," by Bill Niman and Nicolette Hahn Niman, Comments to USDA in Denver, Colorado, January 16, 2007, available at: http://www.ams.usda.gov/AMSv1.0/getfile?dDocName=STELPRDC5058750.

Chapter 11: Answering Obstacles to Reform

238 *Michigan State University Survey:* Conner, D., "The Prospects for Pasture-based Agriculture in Michigan, Overview of Findings," Michigan State University, July 2007, available at: http://www.mottgroup.msu.edu/Portals/mottgroup/PastureBasedAgOverview.pdf.

239 *"[S]mall family and part-time farms are at least as efficient . . .":* Peterson, W., "Are Large Farms More Efficient?" University of Minnesota Staff Paper Series, Department of Applied Economics, College of Agriculture, Food and Environmental Sciences, p. 13 (1997).

239 *"size has little to do with efficiency,":* Thu, K. & Durrenberger, P., *Pigs, Profits and Rural Communities,* State University of New York Press p. 10 (1998).

239 *"size of operations explained less than 5 percent . . .":* Thu & Durrenberger, p. 10.

239 *records of ninety-one Kansas hog operations:* Langemeier, M. & Schroeder, T., "Economies of Size for Farrow-To-Finish Hog Production in Kansas," *Swine Day* (1993).

240 *costs were 5 to 10 percent lower for the free-range farms:* Leopold Center for Sustainable Agriculture, *Swine System Options for Iowa* (May 1996).

240 *machinery, production, and feed for confinement systems made the pasture farms more economically viable:* Frank, G., Klemme, R., et al., "Economics of Alternative Dairy Grazing Scenarios," vol. 28, no. 3, *Managing Agricultural Resources* (October 1995).

240 *grazing dairy farmers made more money per animal:* Winsten, J., et al, *Economics of Feeding Dairy Cows on Well-Managed Pastures,* available at: http://pss.uvm.edu/vtcrops/research/pasture/Economics.html.

240 *"[I]f all the economic costs of [industrial operations] are considered . . .":* Weida, W., *Concentrated Animal Feeding Operations and the Economics of Efficiency,* Department of Economics, The Colorado College, p. 9, March 19, 2000.

241 *"Animal manure is costly to move . . .":* McBride, W. & Key, N., *Economic and Structural Relationships in U.S. Hog Production,* a report of U.S. Department of Agriculture / Economic Research Service, pp. 6,13, (February 2003).

242 *controlled by the top four agribusiness giants of the respective sectors:* Website of Organization for Competitive Markets, www.competitivemarkets.com , (accessed July 2007).

242 *almost all of the rest are produced for those corporations under contract:* Martinez, S. "A Comparison of Vertical Coordination in the U.S. Poultry, Egg and Pork Industries," USDA publication, (May 2002), available at: http://151.121.68.30/publications/aib747/aib74705.pdf.

242 *"A producer without meaningful competitive options . . .":* Harl, N., "Economic Impact and Impacts of Continuing to Proceed as We Are Now," Presented at conference: "Concentration in Agriculture: How Much, How Serious, and Why Worry?", Iowa State University, Ames, Iowa, p.1 (February 2003).

242 *enforcement is seriously deficient throughout much of Midwest farm territory:* Going to Market. *The Cost of Industrialized Agriculture,* Report of the Izaak Walton League (January 2002).

243 *In 2005, Americans paid agriculture more than $21 billion:* "Total USDA Subsidies in the United States," Website of Environmental Working Group, available at: http://farm.ewg.org/farm/progdetail. php?fips=00000&progcode=*total.*

243 *84 percent of federal farm subsidies went to the largest:* Statement based on the analysis of the nonprofit organization Environmental Working Group for the years 2003 to 2005. Data available at: http://farm.ewg.org/sites/farmbill2007/progdetail1614.php?fips=00000&progcode=farmprog&page=conc.

243 *feed is the single largest cost of raising animals:* Wise, T., *Identifying the Real Winners from U.S. Agricultural Policy,* Tufts University, December 2005, available at: http://www.nffc.net/resources/reports/05-07RealWinnersUSAg.pdf.

243 *Feed costs account for 46 percent of total costs in the dairy industry: Farm Economics: Facts and Opinions,* University of Illinois, October 20, 2006, available at: http://www.farmdoc.uiuc.edu/manage/newsletters/fefo06_17/fefo06_17.html.

243 *Dr. Timothy Wise . . . has closely examined:* Wise, T., 2005.

244 *agribusiness regularly seeks and is granted other forms of public assistance:* Lawrence, p. 9.

244 *"the mid-range cost for waste treatment alone . . .":* Weida, W., *Economic Implications of Confined Animal Feeding Operations,* Department of Economics, The Colorado College, January 5, 2000.

244 *"The fact that [industrial animal operations] try to avoid this cost . . .":* Weida, March 19, p. 8.

245 *industrial hog facilities actually harm rural economies:* Gomez, M. & Zhang, L., *Impacts of Hog Concentration on Economic Growth in Rural Illinois: An Econometric Analysis,* Illinois State University, Presented to the American Agricultural Economics Association Annual Meeting in Tampa (August 2000).

245 *industrial agriculture fails to support local economies:* Chism, J., and Levins, R., "Farm Spending and Local Selling: How do They Match Up?" *Minnesota Agricultural Economist,* 676, St. Paul: University of Minnesota Extension Service (Spring 1994).

246 *"[l]arge-scale, specialized operations produce more hogs per person employed . . .":* Ikerd, J., *Economic Impacts of Increased Contract Swine Production in Missouri, Another Viewpoint,* in Livestock Production for Sustainable Rural Communities, Center for Rural Affairs and North Central Rural Development Center, Kansas City, MO. p. 2 October 28-30 (1994).

246 *eight seconds per day in the company of a human:* From McBride & Key, USDA / ERS, p. 14, table 3 (2003).

248 *100,000 human infections and nearly 19,000 deaths from MRSA:* Website of IATP Health Observatory, citing Klevens et al. 2007, posting from November 6, 2007, available at: http://www.healthobservatory.org/headlines.cfm?refid=100687.

248 *"Farm animal MRSA is spreading like wildfire on intensive farms . . .":* Website of the UK Soil Association, citing letter to Dutch Parliament from Dr. C.P. Veerman, Dutch Minister of Agriculture, Nature and Food Standards, December 2006, available at: http://www.soilassociation.org/antibiotics.

248 *The study discovered MRSA at 45 percent of farms:* November 6, 2007, available at: http://www.healthobservatory.org/headlines.cfm?refid=100687.

249 *"Despite all the scare-mongering . . .":* "MRSA Scare Blown Out of Proportion," *National Hog Farmer,* December 15, 2007.

251 *World Society for the Protection of Animals specifically links hunger to industrial animal production:* "WSPA Report Says Factory Farming Exacerbates Global Poverty," PR Newswire, May 23, 2007.

252 *3,800 calories per day of food for every man, woman and child:* Nestle, M., *Food Politics,* Univ. of California Press, p. 13 (2002).

252 *the "greatest unspoken secret" of the U.S. food system is "overabundance.":* Nestle, p. 13.

252 *4.3 pounds of food per person per day:* Kimbrell, A., *The Fatal Harvest Reader: The Tragedy of Industrial Agriculture,* Island Press, p. 7 (2002).

252 *there is now more food per person available on the planet than ever before in history:* Kimbrell, p. 7.

252–53 *"Hunger . . . is a political and social problem . . .":* McLaughlin, M., *World Food Security: A Catholic View of Food Policy in the New Millennium,* Center for Concern (2002).

253 *the height of the 1980s famine Ethiopia continued exporting:* Kimbrell, p. 7.

253 *Large numbers of displaced farmers have joined the ranks of the world's landless poor:* Kimbrell, p. 8.

254 *grazing pigs consume less feed:* Anderson, A., *Swine Management,* J.B. Lippincott Co., p. 220 (1950).

254 *most genetically altered crops are not destined for hungry people:* "Who Benefits from GM Crops?" Report by Friends of the Earth, (2008), available at: http://www.foe.co.uk/resource/briefings/who_benefits_summary.pdf.

255 *taking genes from roundworms and splicing them with genes of pigs:* Kolata., G., "Cloning May Lead to Healthy Pork," *New York Times,* March 27, 2006.

255 *cloning technology is plagued with serious animal welfare problems:* HSUS Letter to Washington Post, February 2, 2008, from Michael Greger, MD, Dir. Public Health and Animal Agriculture, Humane Society of the United States.

255–56 *"Proponents of cloning technology say . . .":* Martin, A., *New York Times,* p. A1, January 16, 2008.

257 *"I think the clearest way to understand the problem with cloning is to consider a broader question . . .":* Klinkenborg, V., "Closing the Barn Door After the Cows Have Gotten Out," *New York Times,* January 23, 2008.

Chapter 12: Back on the Ranch

262 *my public speaking and writing, all of which advocates for returning common sense:* My writings include the following essays in the *New York Times*: "Pig Out," March 14, 2007; "A Load of Manure," March 4, 2006; and "The Unkindest Cut," March 7, 2005.

269 *"[P]asture-based farmers commonly say that their operations do not generate complaints . . ."* Conner, D., "The Prospects for Pasture-based Agriculture in Michigan: Overview of Findings," Michigan State University, July 2007, available at: www.mottgroup.msu.edu.

271 *Americans spend a much smaller portion of their disposable incomes on food:* USDA data, cited in *Salem News,* July 19, 2006, available at: http://salem-news.com/articles/july192006/food_prices_71906.php.

272 *they considered it important that animals raised for their food be treated humanely:* Conner, D., "The Prospects for Pasture-based Agriculture in Michigan: Overview of Findings," Michigan State University (July 2007), available at: www.mottgroup.msu.edu.

272 *81 percent of respondents agreed that the well-being of farm animals is just as important:* Rauch, A. & Sharp, J., *Ohioans Attitudes about Animal Welfare,* Department of Human and Community Resource Development, Ohio State University (January 2005).

272 *animal welfare was more important to them than low meat prices:* Lusk, J., et al., "Consumer Preferences for Farm Animal Welfare: Results of a Nationwide Telephone Survey," Department of Agricultural Economics, Oklahoma State University, August 17, 2007.

273 *The Klessig family creamery:* www.saxoncreamery.com.

273 *Today, there are more than 4,400:* USDA data, available at: http://www.ams.usda.gov/farmersmarkets/FarmersMarketGrowth.htm.

273 *organic farming is "one of the fastest growing sectors of U.S. agriculture . . .":* Thilmany, D., "The US Organic Industry: Important Trends and Emerging Issues for the USDA," Colorado State University, (citing, Oberholtzer) (April 2006), available at: http://organic.colostate.edu/research_docs/Thilmany_paper.pdf.

275 *"May all beings be free from enmity . . ."* Swearer, D., "Buddhism and Ecology: Challenge and Promise," Harvard Divinity School (1998), available at: http://environment.harvard.edu/religion/religion/buddhism/index.html.

275 *Mohammed taught that there is a reward in doing good to every living thing:* Hossein Nasr, S., "Islam, the Contemporary Islamic World, and the Environmental Crisis," *Islam and Ecology,* Harvard Univ. Press, p. 97 (2003).

276 *A* hadith *of Mohammed:* The *hadiths* are the oral traditions about the words and deeds of the Prophet Mohammed.

276 *"If without reason anyone kills a sparrow . . ."* Ozdemir, I., "Environmental Ethics from a Qur'anic Perspective," *Islam and Ecology,* p. 35 (2003).

276 *Jewish laws repeatedly command good treatment of animals:* Goodman, L., "Respect for Nature in the Jewish Tradition," *Judaism and Ecology,* Harvard Univ. Press, p. 251 (2002).

276 *one has a religious obligation to help push the cart lest the owner beat the animals:* Code 191.2. Goodman, L., "Respect for Nature in the Jewish Tradition," *Judaism and Ecology,* Harvard Univ. Press, p. 251 (2002).

276 *"[A]sk the beasts, and they will teach you . . ."* From: Job 12: 7–10.

276 *"For every animal of the forest is mine . . ."* From: Psalms 50: 10–11.

276 *"Be praised my Lord with all your creatures . . ."* From: The Canticle of Brother Sun, by St. Francis of Assisi.

276 *"Are not five sparrows sold for two pennies?"* From: Luke 12:6.

277 *"We support a sustainable agricultural system . . ."* Official Statement of the United Methodist Church, (Paragraph 160 O, sustainable agriculture, Social Principles), available at: http://www.hsus.org/religion/profiles/.

INDEX